A Chronicle and Flora of Niihau

A Chronicle and Flora of Niihau

by
Juliet Rice Wichman
and
Harold St. John

National Tropical Botanical Garden
P. O. Box 340, Lawai, Kauai, Hawaii, 96765, USA

QK
473
.H4
W52
1990

©1990 by National Tropical Botanical Garden.
All rights reserved.
ISBN 0-915809-14-1

Printed in Taiwan by China Color Printing Co.

in honor of

Aylmer Francis Robinson

Table of Contents

Foreward   ix
A Chronicle of Niihau   1
    Glossary of Hawaiian Terms   35
    Select Bibliography   36
A Flora of Niihau Island   37
    Geography   37
    Climate   38
    History   38
    Explorers   39
    Botanical Exploration   39
    Vegetation   43
    Drift Seed   45
    Crops Cultivated by the Early Hawaiians   45
    Catalog of the Flora   46
        Algae   46
        Fungi   47
        Lichens   48
        Ferns and Fern Allies   51
        Conifers   53
        Flowering Plants   53
        Select Bibliography   148
        Index   152

Foreward

This book represents the special accomplishments of two individuals who have long been associated with the National Tropical Botanical Garden. Mrs. Juliet Rice Wichman (1901-1987) was a trustee of the Garden, noted author, and native of Kauai. Her entire life was devoted to the history of our Islands and her love of plants. Dr. Harold St. John (born 1892) is one of our islands best known botanists and has devoted much of his life to the study of our islands' flora.

Dr. St. John's flora of Niihau was completed a number of years ago and submitted to the Garden for publication. Upon hearing of this work, Mrs. Wichman undertook the preparation of her chronicle of Niihau, which was to help give the flora a historical perspective. We are now happy to see both of these works in print and combined in a joint publication.

Special thanks are due to Mr. John H.R. Plews, trustee of the Garden and nephew of Mrs. Wichman, for his diligence in seeing that these works were completed and edited for publication. Mr. Plews carefully checked Mrs. Wichman's and Dr. St. John's manuscripts and helped edit them. He is responsible for the index of Hawaiian names for the flora, select bibliography for the chronicle, and the checking of many references, etc. Gregory A. Koob, Assistant Director—Living Collections, did the typesetting with the assistance of Janet Leopold.

My special thanks to all of the above for their participation in the project, good will, and many helpful ideas and suggestions.

William L. Theobald
Director, National Tropical
Botanical Garden

A Chronicle of Niihau

by

Juliet Rice Wichman

Kamehameha V by the grace of god, king of the Hawaiian islands, by his royal patent, makes known unto all men, that he has, for himself and his successors in office, this day granted and given in fee simple, unto James McHutcheson Sinclair and Francis Sinclair as tenants in common, for the consideration of ten thousand dollars paid into the royal exchanger, the whole of the lands now belonging to the government on the island of Niihau...

Thus, on January 23, 1864, by Royal Patent 2944 the greater part of the island of Niihau came into the possession of the brothers James and Francis Sinclair, who very shortly afterward, on February 13, 1864, added to their holdings the two remaining *ahupuaa* of Kahuku and Halawela which had previously been granted (Royal Patent 5573) to the *konohiki* chief, Koakanu.

A fifty acre *kuleana* held by Papapa (Royal Patent 1615) was not acquired by the Sinclair family until 1880, at which time all of the island with the exception of the school site, the church site, a boat landing, and "all mineral and metallic mines of every description" were united under this single ownership.

There is probably no more romantic saga of the European family in nineteenth century Polynesia than that of the Sinclairs, but before their arrival in Honolulu in 1863, Niihau had had over eighty years of European "discovery", use and misuse by an assortment of British explorers and, later, American sea captains.

Niihau, smallest inhabited island of the Hawaiian archipelago, lies to the southwest of larger and more densely populated Kauai.

It is a dry island, with no fresh water streams and only brackish wells. There is one village, with a church and a public school—no post office, no store. Approximately 41 families live there, of mainly Hawaiian blood.

Politically, Niihau is now a part of the County of Kauai, with 166 registered voters. When in 1821 Kaumualii, the last independent king of the two islands, was abducted and taken to Oahu by Kaahumanu, the widow of Kamehameha I, Niihau and Kauai were absorbed into the Hawaiian Kingdom of the Kamehameha dynasty.

The ruling chiefs then, as do the governmental officers now, lived on Kauai. The closest ties existed between Waimea and Niihau. Communication and barter in the earliest period was by canoe. Poi, the starch staple of the Hawaiian diet, made from the cooked *taro* root, was taken from Waimea to Niihau and exchanged for dried fish and finely woven reed mats. Sometimes drinking water in gourd calabashes was included in the exchange. Waimea had a plentiful supply of fresh water and the Niihau natives produced gourd containers with intricate and sometimes beautiful designs, some of which may still be seen in family collections and in a few local museums.

The earliest written records of Niihau date from the arrival of Captain James Cook in 1778 on his third voyage of discovery in the Pacific. Kauai and Niihau (Oneehow, as the island's name was phonetically transcribed in the early journals) were the first of the Hawaiian islands visited by Cook. He took on a supply of yams (*Dioscorea alata*) and left a "Ram goat and two Ewes, a Boar and a Sow pig of the English breed, the seeds of Millons, Pumpkins and onions."...at Niihau.

It was the yams that brought the expedition back in 1779—they were "the most essential acquisition...as they were a substitute for the Bread which is the most exhausted Article."

The island at that time was described as for the most part bare of trees—with limited patches planted with yams and sweet potatoes, while large plains of fine land are suffered to be waste. Cook procured enough yams however "to serve the 2 Ships for bread for about six weeks."

After the Cook expedition Captains Portlock (1786), Dixon (1787), Colnett (1788), and Mears (1789) visited Niihau, and Captain Vancouver followed in 1792, '93, and '94.

Nathaniel Portlock, (who had been with Cook) commanded the *King George* and *Queen Charlotte* on a "Voyage Round the World". He writes on June 5, 1786, "I determined to proceed to Oneehow without loss of time, in order to get a supply of yams, which I knew that island produced in great plenty and perfection."

On June 8 he records, "Our principal business here was to procure a good stock of yams; and these I had the pleasure to see brought to us in tolerable plenty."

Portlock later "made an excursion into the country" and notes that "The island appears well cultivated; its principal produce is yams. There are besides, sweet potatoes, sugar cane, and the sweet root called *tee* [*Cordyline terminalis*] by natives. A few trees are scattered here and there, but in little order or variety. Some that grew near the well...were about fifteen feet high and proportionately thick; with spreading branches and a smooth bark; the leaves were round, and they bore a kind of nut somewhat resembling our walnut. Another kind were nine feet high and had blossoms of a beautiful pink color. I also noticed another variety, with nuts growing on them like our horse chestnut. These nuts, I understand the inhabitants use as a substitute for candles and they give a most excellent light."

Undoubtedly these trees were the *milo* (*Thespesia populnea*), the *kou* (*Cordia subcordata*)—the wood of both used by the Hawaiians for meat and fish platters, poi bowls, and other containers—and the *kukui* (*Aleurites moluccana*).

Portlock also comments on two wells, one of which he found "brackish and stinking". He does not recommend Niihau as a place to refresh ships' water kegs and comments on the dominance of "Ta'aao", (Kaeo—the Kauai chief) over the island's natives—gifts given to "Abenooe" (who governed locally) being immediately dispatched to the ruling chief in Waimea by canoe. He continues—"the inhabitants at this island are not numerous, and they were kept in such excellent order by Abenooe, that our people walked about wherever inclination led them, without the least molestation."

"Besides hogs and vegetables, we purchased some salt fish of various kinds, such as snappers, rock-cod, and bonetta, all well cured and very fine. The natives likewise brought us water in calabashes, sufficient for daily use, and to replace what had been expended since we left Woahoo [Oahu]. Curiosities too found their way to market, and I purchased two very curious fly flaps, the upper part composed of beautiful variegated feathers; the handles were human bone, inlaid with tortoiseshell in the neatest manner, which gave the appearance of fineered work."

By the time Portlock left he had "procured near ten tons of fine yams" for the *King George* and Captain Dixon had about eight tons on the *Queen Charlotte.*

The Niihau yams, *uhi*, were large, with a black hairy outer covering, white, and mealy within when cooked. Supplies of water, hogs, bananas, and sweet potatoes were more easily obtained for ships' refreshment on Oahu, Maui, Hawaii, and Kauai; but the yams were what brought the ships to Niihau.

Captain James Colnett, who had been with the Cook expedition on the *Resolution*, later became captain of the *Prince of Wales.* He and Charles Duncan, captain of the *Princess Royal* were off Kauai and Niihau during February and March of 1788.

They had (undisclosed) trouble with Kaeo (the Kauai King), and the *konohiki* chief "Abenooe". On Duncan's advice the two ships left immediately for Niihau before the natives there heard of the trouble and hence might not trade.

They succeeded in taking on board yams, wood, water, salt, salt fish, and fresh fish before "Abenooe" arrived from Waimea by canoe and tabued further trading.

In October of 1788 Captain John Meares in the *Felice* recorded: "...We had at this time neither bread or flour on board, and depended on procuring a quantity of yams sufficient to supply our wants during the remainder of the voyage. But as this was not the season for them, and they were too young to be dug up, we should have found it a matter of great difficulty to have obtained a sufficient quantity, if our friend Friday [a Niihau native named Ku] had not undertaken the important negotiation. We, therefore, provided him with such articles as were most likely to forward our purposes; and, by his influence and perseverance, assisted with the bribes in his possession, he persuaded many of his friends to dig up the largest yams they could find, and bring them to market; so that we at length obtained several tons of the most necessary provisions by the morning of the 27th; and at noon we prepared to put to sea."

Meares reports that Captain Douglas in the *Iphigenia* on March 18, 1789, " ...got under way for Oneehow, in order to obtain a supply of yams." Because of adverse winds, they could not make Yam Bay and made for another bay. Here, the quarter-master and two sailors escaped to the island. "Two of them, however, by the active zeal of honest Friday, a native of Oneehow who has already been mentioned...were shortly brought back to the ship; but the quarter-master, who was the ringleader in the mischief, could not be brought off on account of the surf, and was therefore left behind..."

This unnamed quarter-master appears to be the first European resident of Niihau.

The first American vessel to visit Niihau was the *Columbia* under captain Robert Grey in 1789.

Grey took a Niihau boy on board, who became known as Jack Atoo. "Jack" who signed on as a seaman while the ship was in Canton, China, on November 16, 1789, was the first native Hawaiian to reach New England.

He sailed again with Grey returning briefly to Niihau in 1792, where he met with his father, but chose to continue to go on with Grey on the *Columbia*, the first American vessel to sail around the world.

When Captain George Vancouver in the *Discovery* sloop of war accompanied by the armed tender *Chatham* reached Kauai in March of 1792 they were urged "in the strongest terms" to take two chiefs from Waimea when they sailed to Niihau, as "they would be very serviceable in procuring us the different productions, and would prevent any disorderly behavior on the part of the inhabitants."

On their arrival they anchored off the south part of the island and, in spite of the protests from one of the accompanying chiefs that the natives would have a great distance to bring their yams and other food stuffs, they were able to purchase "a very ample supply".

Master's mate Thomas Manby recorded in his journal, "The reason of our touching here was to procure yams, Onehow being famous for the growth of this most excellent root. Some canoes came off immediately on our anchoring and our Traffic commenced for this nutritious branch of commerce.

"By the 16th we had purchased five hundred weight of Yams for Nails and pieces of Iron. Besides a great plenty of Sweet Potatoes.

"They have but few Hogs and trust the produce of the sea for animal food. Being expert and diligent Fishermen they salt a great deal and barter it with the people of Atooe [Kauai] for cloth and mats.

"The fish they take are principally Bonettos, Dolphins, Albicones, and Cavallies of a very large size. An ample supply of this was laid in for our Sea stack making a pleasing variety in our diet.

"I made an excursion with the Botanist nearly round the Island; it is very inferior in point of beauty to all the other of the Sandwich groupe; it produces very few trees of any kind, and is only remarkable for the fineness of its Yams and Sweet Potatoes. I killed a few ducks in my walk, some curlews, and some other birds of a smaller kind.

"Oneehow is considered as under the government of Atooe [Kauai]: it is but thinly inhabited, and had no Chief of consequence residing on it.

"Old Onemoo sent a person of distinction with us from Atooe; he was paid great respect to, and had the sole direction of our Market."

Archibald Menzies who was surgeon and botanist with Vancouver, left an extensive account of this visit on Niihau, which was edited by W.F. Wilson and published in Honolulu by the New Freedom Press under the title *Hawaii Nei 128 Years Ago*.

He speaks of Ku "the same chief that Mr. Meares named Friday" as being most useful because of his authority over the natives and his "obliging disposition".

Menzies went ashore to explore and found "a few small huts, seemingly the temporary residence of a party of fishermen", some platforms for drying fish and some "natives".

Inquiring for fresh water he was shown a place in the rocks where a little water "oozed out by drops" and was convinced by the care that was taken of this small supply that water was a scarce article.

"After examining this romantic bluff...we crossed a low narrow neck of land and pursued our journey along the eastern shore for

Hawaiian canoes greet a visiting ship.

nearly two leagues without seeing anything deserving of notice excepting the desolate and barren appearance of the country we travelled through, covered with loose stones of a black and porous texture and a few stunted vegetables in a shrivelled state—no trees or bushes—no houses or any trace of cultivation were to be seen in the whole tract... On the western shore we saw a few villages and some appearance or cultivation, but in the interior part of the country the same effete appearance prevailed. We passed indeed some small fields of sweet potatoes, which the natives were obliged to cover over with a layer of grass to preserve the little moisture of the soil from being exhaled by the sun's powerful heat. In the middle of the island we saw a large patch of low land encrusted over with salt, which natives told us was overflowed with water in the rainy season, and shows that the soil must be strongly impregnated with that mineral. Though we here and there met with little natural tanks in the rocks which were carefully shaded over with stones to preserve the water that fell in them in the rainy weather, yet these were at this time either dried up or their contents not drinkable, so that for quenching our thirst, we were chiefly indebted to some water melons we obtained from the natives."

They met "numbers of the natives coming loaded with their yams and vegetables from the north end of the island, carrying them to the vessels on their backs at least ten miles in the heat of the day, to obtain a few small nails for the fruits of their laborious industry".

Menzies notes that "musk and watermelons" have been added to the island's produce and "exceeding good savoys". He adds that to encourage the raising of these exotics a good price was always paid for them and that he, Menzies, distributed a variety of European garden seeds—"particularly of Imperial cabbage" to the farmers.

Edward Bell, clerk on the *Chatham*, leaves an account of his exploration of the island in the log of that vessel. Of especial interest in his description (March 1792) is evidence of the coming drought. December, January, and February would usually have been the rainy period when fresh water was stored in rocky catchments.

"...we pass'd some plantations of Yams, & Sugar Cane, which last seem'd in fine condition; but in our whole excursion we did not see a Single Tree, nor a drop of fresh water—here and there we observed holes in the Rocks—which appeared to have been made by art for the

A young Hawaiian woman.

Purpose of Catching rain water, to which the Natives as they pass'd eagerly ran, and seem'd much disappointed at not being able to allay their thirst for there was not a drop in any of them."

In November of that same year (1792) Captain Charles Barkley in the log of the *Halcyon* notes, "They say at Oneehau there is a famine and no yams can be procured".

In 1793 Vancouver returned to the "Sandwich Islands" which was the name Captain Cook had given the Hawaiian group to honor his patron the Earl of Sandwich (yes, the same one who had meats laid between slices of bread to facilitate the taking of nourishment at the gaming table)—with every intention of visiting Niihau.

He was told, however, on Owhyhee (Hawaii) that "...the inhabitants of Onehau had almost intirely abandoned it, in consequence of the excessive drought that had prevailed during the last summer."

Menzies wrote, "We were...disappointed when we were told that we could get no yams this year at Niihau, an island which always used to afford such abundant supply of the useful article... A long succession of dry weather had shrivelled and burnt up the greatest part of the produce of that island, so that most of the inhabitants were obliged to leave it..."

However Vancouver did return to Waimea on Kauai to discharge an obligation he had assumed at Nootka the previous October.

There, lying at anchor off the Pacific Northwest, he had been approached by one Captain Baker of the *Jenny*.

It seems that when the *Jenny* had been at Niihau earlier that year, two young females had been smuggled aboard and well hidden until the ship was too far from land to return them.

The *Jenny*, bound for Bristol, was going straight from Nootka to England and her Captain, James Baker, "earnestly requested" Vancouver to permit "these two unfortunate girls to take passage in the *Discovery* to Onehau, the island of their birth and residence; from whence it seems they had been brought, not only very contrary to their wishes and inclinations, but totally without knowledge or consent of their friends and relations..."

Vancouver assented and took the two girls, "Raheina" and "Tymarow", under his wing, treating them as guests aboard ship, and seeing to it that they were well received and established in adjoining houses of their own before he left them on Kauai.

On the voyage from Nootka to Hawaii the *Discovery* had sailed south to visit some of the Spanish settlements on the continental American coast, where, to quote Vancouver again, the young women were afforded "some recompence for the long and tedious voyage they had been compelled to undertake from their native country".

In Monterey, it seems, European dress was acquired for the girls—riding habits—which they adapted to their own needs, using the full skirts "...as much for concealment, as for warmth;"

These sensational garments must have been the envy of every female heart in Waimea when the girls sashayed ashore from the *Discovery* sixteen years before the *Thaddeus* ever reached Hawaii. Never let it be said that it was the American missionaries who first covered the Polynesian Venuses in European clothes!

A year later Vancouver was off Niihau again, having heard that the yam crop had been reestablished, but was disappointed. Two American vessels, the *Washington* and the *Nancy* were at anchor off Niihau when he left.

But, by 1796, Niihau seemed to be recovering from the drought for Captain Charles Bishop reports in February of that year, "a brisk trade was Carried on for vegetables..." and "...procured a Sufficiency of yams, which are by no means Plentiful at this time..."

He was en route to Canton and had hoped to procure enough yams to last the voyage.

His Majesty's sloop *Providence* and her tender, had preceded Bishop to Niihau by a few days—and had taken on yams, potatoes, watermelons, and pumpkins, noting the presence on the island of Hughes, a European who had come from the Australian penal colony at Botany Bay on an American brig, the *Mercury,* from which he deserted at Niihau.

Kauai, during this period, was in turmoil. A state of war existed following the death of Kaeo, who had been king during the years from Cook through Vancouver, over the succession. Kaumualii was the eventual victor. But at the time Broughton in the *Providence* visited the island, a rival chief, Keawe, held Waimea. This unsettled condition may have caused Broughton's expressions of fear for his safety and that of his men, but it could also have stemmed from the growing resentment among the Niihauans over the constant demands made on their food crops by an ever increasing number of foreign ships—demands enforced by the *konohiki* chiefs sent over by the Kauai overlord to ensure the cooperation of the Niihau farmers.

There was no such thing as private property as we know it in the Hawaii of that time. The land and everything it grew were subject to the king and his designated chiefs. As the value of the Niihau crops became apparent, the few nails and other small objects of barter the farmers themselves received grew to larger "gifts" of red cloth, iron hoops, and even gun powder to the rulers who sent "helpful" representatives along to be sure that these "gifts" reached their own outstretched hands.

Peaceful trading and pleasant relationships between native residents and visiting traders came to an abrupt and bitter end when Captain William Robert Broughton visited Niihau again in July of 1796.

"...we anchored in the afternoon at Yam Bay... As I intended to remain here 48 hours, for the purpose of procuring yams, I sent the boat to expedite our purchase on shore after breakfast, with a small tent to secure them from the sun, and three armed marines to protect the articles they might procure, which I thought would be abundant, and the boat remained at a grapnel manned and armed in case of attack,... In the evening I landed and was sorry to find so small a collection and walked about their plantations to satisfy myself of the imaginary scarcity of vegetables was here: willing, therefore, to make it larger, I walked to some of the plantations but was told there was a general scarcity over the island. On my return, I met with a party which had just come from Atooi, and with them Tupararo, the man who was directed to follow us, that he might supply us with provisions... I walked along shore to the South, where the pinnace waited for me about one mile distant. Only one of the natives accompanied me; and I walked unmolested, meeting several of the inhabitants, till I reached the boat, which was further off than I had imagined. As I had visited this island twice before [he had been with Vancouver] and many of the officers had made shooting parties in the interior without any interruption, I had not the least fear for my own safety; but the unhappy event which took place the next day will show my fortunate escape.

[July 30]"In the forenoon I received some yams from an elderly man,... I also received some provisions from Tupararo, who left the ship with a design, as he said, of sending me more. The cutter was ordered to bring whatever he might have to send; and the mate was commanded to go on shore with two marines properly armed, another man to barter, and the boat's crew with a midshipman to remain off at a grapnel. They were stationed so as to assist in case of need, to be upon their guard, and, if anything occurred, to make a signal. They had not been gone an hour when I was acquainted by the officer on deck, that most of the canoes were gone on shore. I therefore gave directions for the boat's signal to be made: it was then eleven o'clock; we saw them strike the tent, and immediately after heard a firing from the boat. As no signal was made, I thought this firing was intended to recall those who were absent; but soon after, the signal was hoisted for the pinnace, when I sent an officer with the marines to their assistance. On the return of one of the boats, I heard with much concern that two marines were killed; and that the mate, with the botanist, who went to barter, had escaped with the greatest difficulty. The pinnace remained on shore to protect the dead bodies from the natives, who seemed anxious to get them, though they were sunk below the surf. As this unhappy transaction took place without the smallest provocation on our part, I consulted with the officers on the measures necessary to be pursued. Their advice was, at all events to prevent the natives from getting the bodies, and for us to proceed to Attoi, where, by getting some of the chiefs in our power, we might oblige them to deliver up Tupararo, and the other principals in these horrid murders: we also thought it necessary to make some examples on the spot. The boats were therefore manned, the directions given that the marines should burn every house, canoe, and plantation within a mile from the beach were the boats were, and should return before sunset. As they went we heard some firing from the pinnace, which occasioned the boats to land without resistance. The natives took care to keep out of reach of the shot; or if they were near, to drop down on seeing the flash, and then to run away. The houses were soon in flames, and the sixteen canoes on the beach were burnt or destroyed. All this time the natives assembled in great numbers, armed with spears, clubs, and *pahoas*; two of them had the ill-fated marines' muskets and accoutrements. As our people advanced they fled, and so prevented any personal atonement for their treachery. In the meanwhile, the bodies were found in about 9 feet water. At 3 P.M. Mr. Mudge returned, when the natives immediately rushed into the water, searching the bodies, and found the grapnel our people had lost in the attack."

Broughton continued his account, "...it was related to me in the following manner, by Mr. Cowley, the mate, and the botanist, Alexander Bishop, who escaped. The mate, on seeing the signal, ordered the boat in, and struck the tent. The marines unfixed their bayonets; and one of them gave the firelock to the botanist, while he put the tent in the bag. At this instant, while they suspected no danger, the botanist was knocked down from behind; and Tupararo run away with the musket, which the botanist had dropped. The marines were served in the same manner. After, they recovered themselves, there was not time to fire the remaining musket, as the natives pressed upon them with the greatest eagerness on their retreat to the surf. The mate reached the boat in safety, and the botanist escaped by stabbing a man in the water who had seized him; but the ill-fated marines, encumbered by their accoutrements, were murdered in the water by the savages. On examining their bodies, one appeared to have received several stabs with his own bayonet; the other who could not swim, had got a violent contusion on his head, and seemed to have drowned. They were taken up perfectly naked, excepting some few fragments of their trousers."

Ships still visited Niihau in spite of the "massacre". In 1802 the *Atahualpa*, an American vessel, visited the island, and a journal kept on that voyage, (Massachusetts Historical Collection, 1st Series V 9, 1804) reports, "They were the most inconsiderate and lively creatures in the world, they laugh and sing for hours and hours together and do not seem to know what the trouble is...

"Their potatoes are large and good, most of them sweet like the Carolina potatoes; there are several kinds of them, one has exactly the colour and taste of the pumpkin, and another kind is a deep red, or purple colour, like a beet."

The writer found "Their tarro is a very rich root" which when baked had the taste and consistency of bread, adding "if it could be obtained at home I am sure I should never eat any more bread."

One more quotation from a ship's captain—this from *A Narrative of a Voyage to the Pacific, in H.M.S. Blossom in the Years, 1825-8* by Frederick William Beechey, (London, Colburn and Bentley, 1831), "On the 31st of May [1826] we took our leave of Woahoo and proceeded to Oneehow, the westernmost island of the Sandwich groupe, famous for its yams, fruit, and mats. This island is the property of the king, and it is necessary, previous to proceeding tither, to make a bargain with the authorities at Woahoo for what may be required, who in that case send an agent to see the agreement strictly fulfilled. On the 1st of June we hauled into a small sandy bay on the western side of the island, the same in which Vancouver anchored when he was there on a visit of a similar nature to our own; I am sorry to say that like him we were disappointed in the expected supplies; not from their scarcity, but in the consequence of the indolence of the natives.

"Oneehow is comparatively low, and, with the exception of the fruit trees, which are carefully cultivated, it is destitute of wood. The soil is too dry to produce taro, &c. but on that account it is well adapted to the growth of yams, which are excellent and of enormous size. There is but one place in this bay where the boat of a man of war can effect a landing with safety when the sea sets into the bay, which is a very common occurrence; this is on its northern shore, behind a small reef of rocks that lies a little way off the beach, and even here it is necessary to guard against sunken rocks...The natives are a darker race of people than those of Woahoo,...the huts were small, low, and hot; the one which we occupied was so close that we were obliged to make a hole in its side to admit the sea breeze.

"We took on board as many yams as the natives could collect before sunset, and then shaped our course for Kamschatka."

Although the American Board of Foreign Missions had established one of its earliest stations at Waimea, Kauai, very soon after the arrival in the Hawaiian Kingdom of the *Thaddeus* (out of Boston) in 1820, Niihau does not appear to have been considered as a mission site until thirteen years later when this report was sent back to headquarters:

"Niihau: This small island lies twenty miles west of Kauai. The want of good water, occasional famines when the people are obliged to leave and the sparse population seem to exclude the hope that a foreign missionary will be comfortably and usefully settled among them. There are on the island 1,079 inhabitants scattered over a sea coast of 40 miles of dry and barren country. Waimea is rather nearer than any of the other stations, though Hanalei is almost equally accessible and the island could be occasionally visited by a missionary from these stations."

Samuel Whitney, stationed at the Waimea Mission since its beginning, sent an annual report of his endeavors in the field to Headquarters. His references to Niihau tell their own story.

[1840]"It is with feelings of deep regret and concern that I often think of a part of our field (the island of Niihau) as almost excluded from my personal labours, and never more so than just now. Owing to the fact that one of the Catholic converts, a woman of extensive family relations and influence, has been at that Island scattering Catholic books and setting up a school among her relatives."

[1841]"This Island, though included in the Station, is but seldom visited, and has not received the attention and labour which the wants of the people demand. It is separated from us by a somewhat dangerous channel of sixteen miles in width, which is seldom passed except by natives in canoes..."

[1842]"In the month of Oct. I spent a week on the Isl. of Niihau, held a protracted meeting, was well attended; but the prospects of the people on that Island are extremely dark. We have 15 schools...besides eight or ten very indifferently kept on the Island of Niihau..."

There is a manuscript by Gorham Gilman in the files of the Hawaiian Historical Society which gives an interesting picture of Niihau on 1845.

He, Samuel Whitney, and Mr. Tobey crossed the channel between Mana and Niihau in a large double canoe.

[August 27]"Long before the morning light broke in the East, we were seated on the large canoe to go down the coast...There were but very few habitations to be seen, and no signs of cultivation, and as we passed a few natives would stand, almost like statues and gaze at us. Mr. Whitney pointed out to us as we sailed by the most western point of the (Inhabited) Islands... At about 7 o'clock A.M. we ran into a little bay and landed at the village of Keaununui if four houses and an old shed are worthy of that name. The people had been informed of our coming and had prepared for us, and we were better provided for than at Nonapapa a fine large hog was dressing and was soon consigned to his bed of heated stones, and covered up with leaves and dirt, a very fine way of cooking. A few fresh fish broiled on the coals—together with the food we brought afforded us a good breakfast... After breakfast we left Mr. W. to conduct his examination &c. while we started off for a stroll over the country."

They (Gilman and Mr. Tobey) commenced the ascent of "...a mountain which stands isolated from the rest of the range. The ascent was very gradual and after an hours walk we stood on its summit which is quite a tract of tableland so level that standing in the center we could not see below the horizon formed by the brink—from this mountain we could overlook an extensive country, but which had but very little of the beautiful. The land being too sandy to grow much, a few patches of sugar cane and yams were to be seen. Descending we walked through the plantations of the people, but everything bore the marks of the drought of which the people were complaining. We called at several of the native houses in pursuit of mats &c. but found but a very few, and for those they asked us exorbitant prices. As there are no large trees found on the Island the houses are consequently small, and most appear very old and uncomfortable, and the dogs and pigs considered they had equal rights in them."

There being as yet no horses on the island, Gilman and Tobey explored further on foot finding little to interest them and the walk fatiguing.

"A grove of coconut trees situated under the shade of the mountains were the first and only trees worthy that name that we met with. We noticed a number of wells dug through the thin strata of coral, or sandstone, and in which was a small quantity of water, green and brackish, and which would be very hard to drink...we came to the remains of an *heiau* or temple one of the latest built and is in good preservation. It was the dwelling place of Kihawahine a powerful Goddess. It was about 150 feet length 10 feet high 50 wide."

By this time (1847) the governance of the Kingdom of Hawaii had become more regularized. There was a Department of the Interior in Honolulu with Tax Assessors, Land Commissioners, and Agents. There were files for the preservation of records and correspondence, and clerks to write and record them.

One letter, translated from the Hawaiian, gives a glimpse of changes in procurement practices for ships' provisioning.

Interior Department
Honolulu Hale
Feby. 2, 1847

Tax Assessor

At Niihau

A vessel will arrive at the island, if the Captain produce papers bearing Government Stamp, this is it.

Permission is granted him to sail to Niihau, to buy food, has not paid money here, that remains with you to make bargain with him.

These are the restrictions

1. He shall not allow any of his sailors to remain on shore before the vessel sails.

2. He shall not take any native from the island to sail with him. These are strictly prohibited. No women allowed to board the vessel.

Should you see the Captain doing any of these things, stop him, otherwise blame will be laid on you, if you do not attend to it.

With respect,
Keoni Ana

The ships now calling for provisions were from the great Pacific whaling fleet, which in the inactive season filled the harbor at Honolulu and the Lahaina (Maui) roadstead with a forest of tall masts. The overflow anchored off Koloa on Kauai, sometimes a few at Waimea, and an occasional whaler went briefly to Niihau in search of supplies.

In 1850 the *Fortunato* stopped at the island. In 1852 the *America* from Edgartown, Massachusetts, picked up a supply of pigs; two years later, the *Europe*, also from Edgartown found a supply of potatoes.

In 1856, the *Levi Starbuck* was able to get "hogs, sweet potatoes and fowls"; and in 1860, the *Navy* listed a cargo of "30 bbls. sweet potatoes, 17 pigs, some chickens".

Onions were sometimes available but no mention is made of yams, that favorite of British sea captains. One wonders if the Niihau farmers had stopped growing them.

Yams had been bartered for a few nails, an occasional iron hoop, some yardage of red cloth—hogs, sweet potatoes, fowl, and onions now were exchanged, for "cheap prints, a few calico shirts and some cotton pantaloons."

Money—money with which taxes could be paid—was seldom part of the bargain.

Even the small and isolated community on Niihau was expected to pay taxes and lease fees, for this was government owned land. The Niihauans complained about having no industry and therefore no money to pay the high fees demanded of them. W.S. Aka, the Government Land Commissioner for Niihau writes the Minister of Interior suggesting a reduction of fees by $500 a year. He also writes to His Highness, Prince Lot Kamehameha (the Minister of the Interior—later King Kamehameha V), "I am sending you the goat skins belonging to the Government on the Island of Niihau, three hundred, by the hands of Maikai, captain of the vessel Keoni Ana, owned by Elelike.

"Of the number of goat skin, for the freight by the vessel, has been agreed upon at one cent each for all of the goat skins..."

J. Wahineaea, a Land Collector, writes the Clerk of the Minister of the Interior from Kauai, "I wish to inform you that I went to Niihau to demand of the natives their rents for the fifth year, the natives said that there was no money to be had belonging to us now, I said, how about the mats, if you have any on hand, let me have them, and I will take them to the King, who will buy (or sell) them, they replied, there are no mats made up now. That was the first time when I went in the month of December. In the month of March, I went again. This is the amount of money received ($21.50) from the hands of land agent of Niihau..."

It becomes obvious that Niihau is not a money maker for the Kingdom. And in January of 1864, S. Spencer of the Interior Department writes Wahineaea, Land Agent at Niihau: "By direction of his Excellency G.N. Robertson, the Minister of Interior, I beg to inform you that the government has sold all of its right in the lands of Niihau to certain foreigners, a man named Sinclair is one, the other being the foreigner that went with you.

"Therefore the Minister of Interior hereby directs that you and your men proceed at once to Niihau, and catch all the goats, for the reason, that if all the goats are not caught before the 1st day of May, then, the goats remaining shall belong to said foreigners. Do not delay, work with all your might."

Who were these foreigners, this "man named Sinclair" for one? In the preceding years, during and after the Great Mahele, or Land Division, many natives of Niihau had petitioned the Government for permission to purchase home sites on the island. With the exception of the two *ahupuaa* of Kahuku and Halawela and Papapa's *kuleana*, the answer seems to have remained *"Aole makemake ke Aupuni e hoolimalima, aole hoi e kuai aku ka aina ma Niihau"* (The government does not want to rent or sell the Lands of Niihau). (Interior Department Bk. 3, p. 161.)

But in January of 1864 two brothers of Scottish ancestry purchased all the remaining lands on the island from the king, Kamehameha V. Their first offer of $6,000 was refused—the Monarchy holding out for a larger sum. Eventually, $10,000 was paid in gold.

These Sinclair brothers, James and Francis, had arrived in Honolulu on September 16, 1863, from New Zealand on the barque *Bessie*, (300 tons) whose master, Captain Thomas Gay, was their brother-in-law. Francis Sinclair was twenty-three and James some years older.

The Reverend Samuel Damon, who was the Seaman's Chaplain at the port of Honolulu at that time, visited the *Bessie* on her arrival and found to his wonder and amazement that there was but one large family on board "with a beautiful old lady at its head (she was sixty-three), books, pictures, work, even a piano and all that could add refinement to a floating home".

The "beautiful old lady", Elizabeth McHutchison Sinclair, was James' and Francis' widowed mother, and the large family accompanying her consisted of her two sons, three daughters, Jean Gay, Helen Robinson and Anne Sinclair (unmarried and in her early twenties), one granddaughter six years old (Eliza Gay), and four grandsons ranging from eleven to two years of age, George Gay, Aubrey Robinson and Charles and Francis Gay.

The barque *Bessie* had been selected and fitted out in New Zealand by Captain Gay for the Sinclair Family's use, and had sailed by way of Tahiti and Hawaii to the Western Coast of the American Continent in a search for new surroundings and more extensive land holdings.

There were unhappinesses left behind at the old home at Pigeon Bay on New Zealand's South Island. Mrs. Sinclair had lost her husband and eldest son to the sea years before. They had been bound from Akaroa to Wellington by sailing vessel and had never reached Wellington. After a long and fruitless search they were given up for "lost".

Helen (Mrs. Charles Barrington Robinson) had left her husband for undisclosed reasons a year after their marriage, and returned to her mother's home, bringing her infant son, Aubrey, with her—and her brother Francis had recently suffered an unhappy lover's quarrel and broken engagement.

So the prosperous estate at Pigeon Bay had been sold, the *Bessie* fitted with all sort of special comforts for the family: furniture, household linens, books, pictures, jams and jellies, quantities of fresh apples, a milk cow, poultry, Merino sheep, and the family piano.

(An unconfirmed but widely believed tale, which has persisted to this day, is that all the money from the Pigeon Bay estate was carried in gold coin in a stout chest deep in the *Bessie's* hold).

The arrival of the Sinclair family was widely welcomed in Honolulu by the flourishing British community. The King, Kamehameha IV, and his beautiful Queen Emma, both ardently pro-British, received them cordially and encouraged their search for a new home in the Hawaiian kingdom.

Although still a young man, Kamehameha IV died in November of 1863, to be succeeded by his elder brother Lot. Prince Lot, prior to succeeding to the throne, had been Minister of the Interior, hence privy to the Sinclairs interest in the purchase of Niihau. It was from him as Kamehameha V that James and Francis received the royal patent in January 1864.

On February 4, 1864, this quotation from the Pacific Commercial Advertiser of that date was headlined:

Sale of an Island

The Island of Niihau has recently been sold by the Government to the Messrs. Sinclair, for the sum of $10,000 cash. It is the intention of the purchasers to make a sheep range of it, for which the island is said to be well adapted. Niihau is seldom visited by foreigners, on account of its out-of-the-way position, and consequently is among the least known in the group. It is eighteen or twenty miles long, about five or six miles wide, one-half of which consists of elevated or highland, about 800 feet above the level of the sea, and the other a lowland or plain. Both sections are said to be good for grazing. It is on the lowland that the natives, of whom there are perhaps two hundred, live, and support themselves by fishing, and raising sweet potatoes, onions and yams, all of which grow well and are the finest produced on the group. As a sheep range, the island will probably be found to possess some advantages, as all other animals, dogs, etc. can be entirely excluded from it. The few natives who live on the island, by the introduction of wool growing there, will be enabled to assist and have some better means of maintenance than heretofore.

In 1865 William T. Brigham visited Niihau. His Journal in the Kauai Historical Society files has a fascinating account of the trip. William Brigham was a botanist collecting specimens of the Island flora for Harvard University—he later headed the Bishop Museum in Honolulu.

"Tuesday evening I had decided to go to Niihau as a boat was to leave Waimea, and engaged passage accordingly, but as the boatman sent me word that he should not go that night, we went to bed and to sleep. At eleven o'clock a kanaka posted over and told us the boat was waiting, and so I got up and while I was dressing one of the boys caught and saddled my horse, and I rode in the faint moonlight to Waimea. The whale boat was drawn up on the beach and a party of some forty natives was scattered around on the sand; some of the men were asleep, others eating, and the whole scene as the moon occasionally looked out of the clouds, reminded one of a smugglers' rendezvous. I sat with them about an hour and at one o'clock we were put into the boat and it was floated out a few rods from shore and the lading commenced. *Paiai* and poi, fish, and calabashes were piled in unmercifully, and in the midst of this a wave came in and wet me through. Twenty kanakas then got in with three filthy women and three diseased babies, and then the boat loaded almost to the water edge, started on her way. At eleven the next morning we were off Lehua and there our dangers commenced in earnest. The surf was terrible and the kanakas proposed to turn back to Kauai, but at last decided to go on. Often the rowers jumped overboard to lighten and steady the boat as the great waves came rolling upon us. We were so heavily laden that we could not rise on the waves and were in constant danger of being swamped. I was so seasick that I felt quite unconcerned, and had the boat upset, I should, I believe,...have rolled to the bottom without a struggle. The passage between Niihau and Lehua is not more than three quarters of a mile wide, and very shallow, but bottom being perfectly visible and quite rocky. As we rose on the swell we looked out for rocks, and then when we went down we could see nothing but water and sky, and kept our direction as well as we could. As it was impossible to land we pulled fourteen miles and about four o'clock in the afternoon landed on the southwestern end of the island. I was thankful to have escaped from the most dangerous voyage I had yet made, and as I had neither food nor water for twenty-four hours, lost no time in getting a horse and riding about a mile to the Sinclairs, a Scotch family from New Zealand, who have lately purchased the whole island. I had never seen any of the family before, but notwithstanding this I was received in a most cordial manner, and soon felt at home. Thursday morning with Mr. Sinclair and two of the boys ([eleven year old] Charles Gay and Aubrey Robinson) I rode around the south end of the island. This, as well as the north and west portions of the island, is comparatively level and at some former age was perhaps the sea bottom. Coral reefs and sandstone occur nearly a hundred feet above the sea level, and the sand may be found everywhere by digging from six to ten feet or even less in some places. Several tufa craters are on the coast and near one we found a small spring of fine water. On the beach by this spring were large quantities of drift wood, some from Oregon probably. The tropic birds were very abundant, especially a species I had not seen before (*Phaethon rubricauda*), with red bills and tail feathers. Some calcareous crystals were embedded in the tufa with fragments of coral as at Leahi on Oahu. In the ancient elevated reef which was hard and sonorous, were large cracks and cavities in which grew sugar cane in large quantities and breadfruit trees, the tops of the latter seldom or never rising above the ground level. On the beach I noticed a rock shaped much like a human head, and doubtless it has been the object of many a native prayer.

"Niihau is celebrated for its pineapples and we found that we could eat several in succession and be ready, like Oliver Twist, for more. The nopal cactus is very common and indeed is the largest shrub we saw. Wild cotton was seen in a few places, and also a few guavas, but the mass of vegetation consists of argemone, asclepias, indigo, *ohai*, tephrosia and several smaller plants. The central portion of the island is high and on these hills are small trees which furnish a supply of firewood. A curious lobelia is also found on the cliffs, having leaves like a violet and small green flowers on the end of a very thick stem (*Delissea undulata*). Friday morning at five o'clock I started for the little village of Lehua opposite the islet of the same name, and arrived there at half past six after a ride of about fourteen miles. On the way I passed several fresh water ponds and very few native houses. The Sinclairs had very kindly placed their boat at my

The island of Lehua. *(photo by J.H.R. Plews)*

disposal, and had sent it over to this place the night before. It was a beautiful little boat, and with three rowers and a bailer we were well equipped.

"... In former times the population of Niihau was two or three thousand and the manufacture of fine mats from rush was an important industry; sweet potatoes and yams were also raised to supply the people of Kauai and later the whalers. Now there are some fifty men and a hundred women on the island. The ancient inhabitants constructed with much labor and skill tanks or reservoirs for rain water, some of what are in use at the present time while others are lost. All vegetables grow well although the soil is not very fertile, and the Sinclairs are planting the excellent *manienie* grass in large quantities. A few coconuts constitute the only trees on the island..."

The Church of Niihau reports in the Hawaiian Mission Children's Society Library in 1867 that after the arrival of the Sinclair family

Sundays had become peaceful in contrast to the time when many whale ships had come to the Island, and Sunday was a "day of gadding about".

The writer (this has been translated from the Hawaiian language) comments on the help his church has received from the *"haole"* family who "live truly spiritual lives" and "freely help toward the necessities of the church". They also "attend our services on Sunday...the children of the haoles always attend Sunday School and church."

In July of 1867, according to an account in the Pacific Commercial Advertiser (August 3rd) the Sinclairs received, on the schooner *Nettie*, a consignment of lumber to make "additions to their hospitable mansion. And then the active stalwart natives were busy on shore rolling down the bales of wool and cotton...to be shipped..."

"We visited the residence of the Messrs. Sinclair, and although unknown, and without those stereotyped passports of Society—letters of introduction—we were warmly welcomed and received and entertained in that genial, old fashioned kind of Scotch hospitality that does one's heart good—that hospitality which springs spontaneously from a warm and generous nature, untrammeled by the cold conventional rules of society, or rather, mock society. That welcome, those exhilarating refreshments around emotions that had long slept...

"The situation of the house is very beautiful and commands an extensive view. It is on a long cape-like ridge, on the northwest side of the island, that terminates in the sand dunes referred to. Trees only are wanting to give it that rural air, which no country mansion can ever wear without trees to wanton in the breeze."

In a copy (October 1870) of *The Friend* a "recent visitor to the Island of Niihau" comments: "On the island of Niihau the Sinclairs have, I think, their full heart's desire. I never was more pleasantly disappointed in regard to any place. Viewed from Kauai, it presents a most uninteresting appearance, which is very deceptive. It is about twenty miles long, and five or six wide, containing over sixty thousand acres of land. The greater part affords most excellent pasturage, especially for sheep".

Once again from a church report (1871) the influence of the Scottish family on the life of the native community is revealed:

The Work of F. Sinclair

"The Hawaiians are very grateful for their *haole* chief in several ways. April is the regular month for sheep-shearing. That is the time when 3000 or more sheep are slaughtered; to eat the mutton and throw (the entrails) into the sea for the sharks.

2. To collect the tallow. The remaining tallow from the slaughtered sheep. The people take that tallow for them and everyone has from one to two barrels. They are well supplied with tallow every year.

3. F. Sinclair gave graciously an eighth of the whole island of Niihau as an open country where the horses of the people might run. He gave the posts and boards for the horse pasture from the uplands down to the sea. That is at Keawanui.

4. He gave freely lands for planting sweet potatoes—two planting places—at Puheheke and below Kamalino, perhaps 100 or more acres.

5. To eat; people are to gather fish surrounding Niihau; there is no disputing about fish, like the day of the "konohiki".

6. He will punish the persons indulging in sexual gratifications and drinking liquor. The Government will punish persons outside of their house lots. He will punish persons who drink liquor inside of all the houses, if discovered.

7. He and his family go regularly to the church services and urges the Hawaiians to come to church. He marks the children that do not go to Sunday School and punishes them if they always stay home. Therefore his Sunday School is full throughout the year. The Sunday School work is progressing finely on Niihau under the excellent leadership of F. Sinclair.

There are very few haole here in Hawaii who do like this, for the two goods—physical and spiritual."

A. Kaukau

By 1865 the family was comfortably settled in the large rambling house on the bluff near Nonopapa. They had bought horses and cattle from already established ranchers on the other islands of the Hawaiian group to add to their flock of fine Merino sheep.

They had purchased a whaleboat for their own use to cross the rough, sometimes very rough, channel between Waimea on Kauai and Kii on Niihau, and manned it with Hawaiian oarsmen. Since none of the three hundred natives living on the island spoke English, the Sinclairs set about mastering the local language in order to become independent of interpreters.

Captain Gay, having seen the family established on Niihau, sailed for Australia where he sold the *Bessie*, and shortly thereafter died in Sydney of pneumonia. He never saw his last child, a daughter Alice, born in March of 1865.

Also in 1865, Francis Sinclair returned to New Zealand to marry a McHutchison cousin. Isabella McHutchison is the Mrs. Francis Sinclair responsible for the book *Indigenous Flora of the Hawaiian Islands* which was published in London in 1885. This volume of reproductions of her exceptional botanical paintings was the first of its kind and is now a treasured and indispensable part of any collection of nineteenth century Hawaiiana.

On the island, tutors for the growing children came and went. Friends from Honolulu and Kauai visited from time to time. An occasional British naval vessel anchored offshore and the officers made "The House" their headquarters. A Lord and Lady Brassey, cruising around the world in their yacht, stopped briefly.

One of their most frequent guests was Valdemar Knudsen from Kekaha on Kauai. Originally from Norway, he had been on Kauai for several years prior to the Sinclair purchase of Niihau.

Fluent in the Hawaiian language, he had been most helpful in many ways, and it was he who eventually was able to persuade Papapa and his wife to sell their *kuleana*, thus giving the Sinclairs sole ownership of the island.

In February of 1867, wearing "a simple white dress and orange blossoms" Anne, the youngest of the three Sinclair sisters, was married to Valdemar Knudsen in the little village church. The next day she left Niihau with her husband to make her home on Kauai.

In that year Francis Sinclair took over the office of District Judge for Niihau from his brother James. James was in poor health due to an accident he had suffered before leaving New Zealand.

In the period between 1867 and 1873, Elizabeth McHutchison Sinclair bought for herself and her daughters Jane Gay and Helen Robinson after her, a large tract of land at Makaweli and Hanapepe on Kauai and began building a home high on the hills overlooking what would one day be a sweep of sugar cane fields.

It was at Makaweli that James Sinclair died in 1873. There is no record of James' will on file, and although *The Hawaiian Annual* of 1889 states that Francis Sinclair is sole owner of Niihau it may well be that James' mother or two sisters Jane Gay and Helen Robinson participated in the distribution of his estate. (Personal communication from a family member.)

In a combined tax report (James and Francis Sinclair) the 1873 value of the land on Niihau is given as $18,000 and the personal property a value of $40,000.

In 1873, ten years after the arrival of the *Bessie* in Honolulu, Mrs. Robinson (Helen) and her son Aubrey, with his cousin Charles Gay, left for Boston where the young men were to pursue their education.

Francis Sinclair traveled extensively in the following years, at the time when his nephew Aubrey was in Boston and after Aubrey's graduation from the Boston University law school in 1875.

Aubrey Robinson, according to *Builders of Hawaii* published by the Honolulu Star Bulletin in 1925, also spent some years after his graduation in travel in England and "on the continent" as well as in the Middle East and the Orient. On his return to the Islands in 1884 he imported the first pure bred Arabian horses from Arabia.

(My mother had a half Arabian mare which she named, suitably enough, Niihau. It had been given her by the Robinsons. I never remember my mother riding any other animal. The mare, Niihau, lived to a ripe old age and was spirited to the end.)

In June of 1885 Aubrey Robinson married his cousin Alice Gay, an event which would later lead to sole ownership of Niihau passing by name, if not by family, from Sinclair to Robinson, under which name it is held today by the widow and two sons of Lester, the youngest of the five children of Alice and Aubrey Robinson.

Niihau's day in the sun as a romantic (to the outside world) and self-contained family home had ended when the matriarch, Elizabeth Sinclair, established herself, her daughters Helen Robinson and Jane Gay, and their children, on the Makaweli, Kauai estate.

On January 31, 1891, after more than twenty-five years of ownership, Francis Sinclair sold to his sisters Jane Gay and Helen Robinson and to his Nephew Aubrey Robinson all of his lands and personal property on Niihau. The sisters paid him $35,000 and Aubrey $1. He was also to be paid on the 31st of December of each year, commencing December 31st, 1891, the sum of $3,500 in U.S. gold coin during his life or for twenty-eight years, when payments would cease.

At approximately this time, Elizabeth Sinclair gave her two daughters "all of my lands and all of my real estate situated on the islands of Kauai and of Niihau".

Mrs. Sinclair, (Elizabeth McHutchison) died in 1892, at the age of 83. In her will, a simple straight forward document, she states that she has already provided all her children with "freehold estates", and gives, now, "all the property of whatsoever nature I may die possessed in equal shares to my two daughters Jane R. Gay and Helen McH. Robinson, their heirs and assigns, with the exception of the one third of my yearly income from Niihau which I have given to my son Francis Sinclair to be enjoyed by him during his life, and $5000...to my daughter Annie McH. Knudsen."

The witnesses to her will were her son Francis Sinclair and grandsons Francis Gay and Aubrey Robinson.

Francis Sinclair lived in England for many years after leaving Niihau. He died at the age of eighty-three on the Isle of Jersey in 1916.

An obituary stated: "During his later years Mr. Sinclair resided in London, devoting himself to literary work, and publishing his *Ballads and Poems from the Pacific*, also several volumes of short stories *Under Western Skies*, *From the Four Winds* and other interesting books".

The House on Niihau was now no longer a family home—it was used, yes, but only occasionally, as a vacation or holiday retreat—on an island that was once more becoming an isolated community of native Hawaiians.

A relevant description of this period can be found in the *Directory and Reference Book of the Hawaiian Islands* (B.L. Finney).

"Its entire land belongs to the firm of Gay and Robinson, who use it for a sheep ranch. The population is not much more than one hundred, mostly men who work on the ranch, doing some fishing between busy seasons, together with their families...

"Nonapapa, on the west side, is the main landing place, where sheep are sheared and the wool is shipped. It is two miles from the manager's residence or chief ranch house. There is a flourishing grove of coconut palms a stone's throw from the landing. Last year thirty thousand sheep were brought to the shearing, and this year it is expected there will be thirty-six thousand. About a thousand head of cattle graze upon the island and the proprietors keep fifty horses there. They are a superior class of horses, too, bred from imported Arabian sires. Two miles to the southward of Nonapapa is a fishing village. It has a schoolhouse where Sunday School is held and a native preacher proclaims the gospel to a congregation of thirty or forty people. Around the church dwellings are grouped, some frame and some grass in construction. There are grass sheds for sheltering boats, which contain a number of boats left by whaling and sealing craft. The inhabitants may be seen drying fish on the beach for shipment to Honolulu. They also ship goatskins to market...The native settlers work for the ranch at shearing time. One of them is the boss shepherd, another the boss cattleman, with lunas under them. They get credit for a full day's work for every distinct task allotted to them...

"On the side of a hill there is a cavity fifteen feet deep, in which three breadfruit trees are growing, their foliage overshadowing the whole area. We took luncheon under this bounteous shade.

"The manager's residence...is two miles from Nonapapa. It is fifteen miles by the road from Ki. Mr. G.N. Moore is the manager, whom I found an agreeable and hospitable entertainer. The residence faces the southwest and consists of nine cottages fronted with a continuous veranda. Part of the barracks is reserved, well furnished and kept, for the accommodation of Messrs. Gay and Robinson, with their families...The demesne occupies an elevated plateau, about ten acres in extent, and is enclosed by a stone wall. Wild turkeys and peacocks, as well as cats, are abundant about the domicile. Water for household purposes is obtained exclusively from the rains."

Aubrey Robinson was his mother's heir with a full half interest in the, at first, informal partnership of Gay and Robinson. As, one after the other, the Gay children married or moved away, lands were bought for them or their inheritance given in some other form. Only Alice was eventually left to share the Gay and Robinson estate with her husband-cousin.

I was fortunate enough, as an eleven year old in 1913, to spend several happy weeks on Niihau with my parents and sister as guests of the Robinson family. We rode horseback from one end of the island to the other, fished, gathered shells and played with two piglets, Snookums and Voice. Lester Robinson was my contemporary and the pet pigs were ours. My sister Edith and Lester's older sister, Eleanor, scorned our childish pleasure in these little animals.

Many years later, in 1963, as the Kauai member of the first State Board of Education I returned for a brief one day visit to the island. We crossed the channel between Kauai and Niihau in a converted W.W. II landing craft that was used for the transportation of supplies and Niihau residents and their families to and from Kauai.

The native flora was still abundant on the beach and the *makaloa* (*Cyperus laevigatus*) reed flourished in damp inland areas. *Kiawe* groves (*Prosopis pallida*) were now widespread and plumbago (*Plumbago capensis*) vined through them showing pale blue flowers. The prickly pears, (*Opuntia megacantha*) that had grown everywhere on my first visit fifty years earlier, were gone. I remembered, however, the cool sweet taste of this fruit which Mr. Robinson had picked and peeled for us with his leather-gloved hands.

Aylmer, the second of the Aubrey Robinson sons, was our host on the trip. After his father, he had become manager, major owner (with his brother Lester) and guiding spirit of the island until his death in 1967. It is to him that this chronicle and flora is dedicated.

A prickly pear cactus (*Opuntia megacantha*) felled by Fusarium blight (a fungal disease), Apana Valley, 1947.
(photo by H. St. John)

Aylmer Francis Robinson was born at Makaweli, Kauai on May 6, 1888. He was a graduate of Harvard University (A.B. 1910). On returning to Hawaii he began work with the Oahu Sugar Company. A year later, in 1912, he became manager of the Gay and Robinson ranch on Kauai and ten years later resigned from that position to become manager of Niihau for his father. He was admitted as a partner into the family firm, Gay and Robinson, and in 1916, four years later, was appointed as its business manager. He never married.

Aylmer inherited a three-fourths interest in Niihau, and Lester a one-fourth interest, on the deaths of their parents, Aubrey and Alice Gay Robinson. In turn, at his death in 1967, Aylmer Robinson left his entire interest in Niihau to his brother Lester and Lester's wife Helen M. Robinson. At Lester's death his widow and their two sons became the owners of the island.

During the years of Aylmer Robinson's management the feral goats on Niihau were eradicated in a major effort to preserve the native flora.

He was host to a gubernatorial party in 1938, an excellent account of which and of the condition of the island at that time appeared in the October 22, 1939, Honolulu Star Bulletin.

On December 7, 1941, a Japanese plane crashed on the island and the armed enemy pilot attempted to take control of the island, but was frustrated and killed by an angered Hawaiian couple.

During the period of martial law which followed the attack on Pearl Harbor, David Larsen, then manager of Makaweli Sugar Co. visited Niihau and from June 17th, 1942, to June 24th wrote his wife Katherine a very long, day by day account of his stay, which may be found in the State Archives, Stacks 916.6 L3r Copy 1. A copy is in the files of the National Tropical Botanical Garden.

Glossary of Hawaiian Terms

ahupua'a	land division, a major chief's estate
haole	foreigner
heiau	temple
kanaka	human being; a Hawaiian
konohiki	overseer of chief's estate
kuleana	homestead or small farm awarded c.1850 to commoners who farmed them
pāhoa	dagger
pa'i'ai	thick taro root paste, when mixed with water becomes poi
uhi	yam, *Dioscorea alata*. "Yam" is the English name for several species of *Dioscorea* with edible, tuberous roots, which look like a large, hairy sweet potato, and inside have a starchy white flesh like a firm, somewhat fibrous white potato. *Uhi* are seldom found in Hawaii's supermarkets, although the Chinese Yam (*Dioscorea batatas*) sometimes is. The word "Yam" is believed to be of African origin, where *Dioscorea* is a crop, and came to be applied in the United States to the sweet potato ('uala or *Ipomoea batatas*) especially those with a moist, sweet, orange flesh.

Select Bibliography

Anon. *Extracts from a Journal kept on board Ship Atahualpa, etc.* Photocopy of pp. 242-245 of original in printed Massachusetts Historical Collection 1st series, vol. 9 (1804) pp. 244-5, in Mission Historical Library.

Barkley, Charles, *Log of the Halcyon*. Photocopy in Bishop Museum. Entry for Nov. 12, 1792.

Beechey, F.W., *Narrative of a Voyage to the Pacific.* (London 1831) pp. 320-1.

[Bell, E], Untitled Journal, photocopy in Hawaiian Historical Society. Published as Log of the Chatham in *Honolulu Mercury* Sep. 1929 p. 24.

Brigham, W.T., Typescript extracts from journal in Kauai Historical Society.

Broughton, W.R., *A Voyage of Discovery to the North Pacific Ocean.* (London 1804) pp. 74-80, Mrs. Wichman's quotation in the text contains a few phrases not in this published version. No copy of the original journal has been found in Hawaii.

Cook, J., *The Journals of Captain James Cook*, Beaglehole, J.C. ed., vol. 3 pp. 276, 573, 1231.

Kaukau, A., Translation [by Rev. Henry Judd], *Parish Report of Niihau 1871* of *Ka Hoike Kihapai o Niihau*, both mss. in Hawaiian Mission Children's Society.

Manby, T., "Journal of Vancouver's Voyage to the Pacific Ocean (1791-1793)", *Honolulu Mercury* July 1929 p. 33.

Meares, J., *Voyages Made in the Years 1788 and 1789.* (London 1790) pp. 281, 357-8.

Menzies, A., *Hawaii Nei 128 Years Ago.* (1920) p. 40.

Portlock, N., *A Voyage Round the World.* (London 1789) pp. 75, 83-90.

Vancouver, G., *The Voyage of George Vancouver.* (Hakhayt Society 1984) pp. 470, 475.

[Whitney, S.W.], Untitled answers to Questions 1, 2 and 3 of ABCFM, 1833-1834. Ms. in Hawaiian Mission Children's Society.

[Whitney, S.W.], Untitled mission station reports for Waimea, Kauai for 1840, 1841 and 1842. Not all signed.

Flora of Niihau Island

by

Harold St. John

Geography

Niihau Island, of the Hawaiian Islands, lies 17.5 miles southwest of the island of Kauai, and 148 mile northwest of Oahu. It is approximately 18 miles long, 6 miles wide, covers an area of about 72 square miles, and is 1,281 feet high at its highest.

The central and eastern part of the island is an upland remnant of a volcanic dome of basaltic lavas, which on the east forms abrupt sea cliffs and on the west slopes down to a coastal plain. Numerous deep valleys cut into the western and southern slopes of this upland remnant island and at the mouths of these valleys playas, or temporary ponds, have developed. Some of these ponds reach considerable size when there is an infrequent, heavy rain and they soon dry to form alkaline flats. The largest, Halalii Lake, is 2½ miles long.

The northern sixth of the island, north of Kaali Cliff, is a lowland of basaltic rocks and calcareous sand dunes with consolidated older dunes and unconsolidated recent ones near the shore. The southern third has a small volcano at Kewaewae, but is largely a low flat of basaltic rocks, partly covered with strips of sand dunes. The southernmost point, Kawaihoa, is another volcanic cone, 548 feet high, of bedded tuff and breccia. More details of the geography and geology can be found by consulting Stearns and Macdonald (1947).

Climate

Niihau is considered tropical, but arid. It lies exactly in the lee of the island of Kauai of which the latter has a large, broad mountain, culminating at Kawaikini, 5,170 feet high. This broad mountain traps nearly all the water from the clouds blown in by the normal northeast tradewinds, which continue for most of the year. Niihau, being much lower, and in the lee of this larger island, never gets any rain from these tradewinds and as a result the island is quite dry most of the year.

Southern, or Kona, storms are infrequent, but occur mostly in the winter. However, they do bring rain, and from 1919 to 1925, the average rainfall was 26.27 inches. Prolonged droughts are characteristic. There are no perennial streams, and the shallow outwash lakes or playas evaporate and are dry most of the year. Wells are few and mostly brackish. Springs do occur, but there are only two permanent ones, the best being 750 feet up the face of Kaali Cliff. Full weather records are not known to be available. However, to a visitor from Honolulu, the climate is hot, the sun glaring, and it is a fine place to get a tan or a sunburn.

History

Niihau has long been inhabited by Hawaiians, and it is estimated that the early population was some 300. This has since dwindled, but is still mostly pure blooded, native Hawaiians.

The island was purchased in 1864 by James and Francis Sinclair and since then, it has been continuously owned by branches of the same family and it is still administered as a sheep and cattle ranch. Also, goats were introduced by Captain James Cook in 1778 and their descendants and the sheep caused deforestation of the uplands which resulted in severe soil erosion in the 19th Century. Elimination of the goats, tree planting, and conservation have restored some forest growth, and introduced grasses have partially restored the plant cover.

Explorers

Numerous early voyagers touched Niihau in search of food supplies and water. Of water, there was little to be had and of foods, foremost was the yam, but there was also sweet potatoes and pigs. Their journals mentioned these crops, but little else of botanical interest. The explorers included Cook in 1778, Dixon in 1787, Colnett in 1788, Portlock in 1789, Vancouver in 1792, Broughton in 1796, Townsend in 1798, Turnbull between 1800 and 1804, Beechey in 1826, and others.

Botanical Exploration

Captain George Vancouver visited Niihau from March 14th to 16th, 1792, and, in January or February, 1794, Dr. Archibald Menzies, his surgeon and active botanist, doubtless collected plants, but because of a dispute between the two men, the records have vanished. His specimens from the island were all labeled merely "Sandwich Islands".

Captain F.W. Beechey on the *Blossom* visited Niihau in 1826 and on his staff were two botanists, George T. Lay and A. Collie. They collected eleven species, two of them weeds; a meager total for two botanists who were on shore for two days.

Jules Rémy spent from 1851 to 1855 in the Hawaiian Islands, travelled to all the major islands, including Kahoolawe and Niihau, and collected plants. He published on history and anthropology, but nothing on botany. His plants specimens were deposited in the Museum d'Histoire Naturelle, Paris, where they were determined, arranged taxonomically, and then numbered. A list of his 726 plants was compiled and the present writer has a copy. For each the only locality data given was the name of the island. The first set was kept in Paris, and one set of duplicates was sent to Asa Gray at Harvard University. Gray detected and described several novelties among them. A duplicate set is also now in the Bishop Museum.

In the Paris collection there are five species recorded for "Nihau", and thirty-five from "Kauai ou Nihau". Of the latter, most are well known on Kauai, but four are endemic to Niihau, and three occur on both Niihau and Kauai. Fifteen are unknown in Niihau, unlikely to have been there, but are well known on Kauai.

Plants Recorded for Niihau by Rémy

257. *Compositae* = *Lipochaeta succulenta* (Hook. & Arn.) DC.
315 bis. *Scaevola* = *S. coriacea* Nutt. The only record for the island, but a credible one.
407. *Vitex* = *V. ovata* Thunb.
416. *Batatas littoralis* = *Ipomoea stolonifera* (Cyrill.) J.F. Gmel. This is still present on the beaches.
534. *Isodendrion* = the holotype of *I. remyi* St. John. It is now extinct.

Plants Recorded for Kauai or Niihau by Rémy

These totaled 35 species and through research in the Paris herbarium, the author has located most of these. The following are endemic to Niihau:

309 ter. *Isotoma* = *Brighamia insignis* A. Gray. St. John (1969: 192) deduced that this was the Niihau species, not the Kauai one.
300 bis. *Delissea niihauensis* St. John. It is now extinct. See St. John (1959: 177).
595. *Euphorbia* = *E. celastroides* var. *stokesii* (Forbes) Sherff. There are two sheets of this. Also, there are three sheets from Kauai of a different looking plant, which the writer determines as var. *humbertii* Sherff, a known Kauai variety.

The following occur both on Kauai and Niihau, so may well have been collected on Niihau:

102 bis. *Echinochloa* = *E. colona* (L.) Link. It is still present.
460. *Lysimachia* = *L. mauritiana* Lam. The duplicate in the Gray Herbarium is labeled "Nihau". It has not been collected there since Rémy's time, but there is no reason to doubt his record.
565. *Thespesia populnea* (L.) Soland. ex Corrêa. It is still present.

Other plants which Rémy marked as from "Kauai ou Nihau", do not now occur on Niihau, and are unlikely ever to have been there, but do actually grow on Kauai. They are:

73. *Selaginella* = *S. arbuscula* (Kaulf.) Spring.
141. *Carex* = *C. meyeniana* Nees. "Oahu, Kauai ou Nihau." There are two sheets of this species, one marked Oahu.
145. *Carex* = *C. wahuensis* var. *rubiginosa* R.W. Krauss. A second sheet is labeled "Hawaii".
156. *Joinvillea* = *J. ascendens* Brongn. & Gris var. *ascendens*. It was listed as from "Hawaii, Maui, Molokai, Kauai ou Nihau, Oahu".
226 bis. *Wikstroemia*. There are two sheets, the first is *W. furcata* (Hbd.) Rock. The second is *W. hanalei* Wawra.
258 bis. *Compositae* = *Bidens laevis* (L.) B.S.P. It is an introduced weed. Rémy's record is the earliest in the islands.
344 bis. *Gardenia taitensis* = *G. remyi* H. Mann.
359. *Psychotria* = *P. mariniana* (C. & S.) Fosb. ? "Molokai, Kauai ou Nihau".
455 bis. *Solanaceae*. Specimen not located.
515. *Broussaisia arguta* Gaud. It is correctly named. Marked "Oahu, Kauai ou Nihau". There are four sheets. It is common on Kauai.
548. *Schiedea* = Holotype of *S. amplexicaulis* H. Mann. Rare, and probably extinct, but once collected on Kauai by Lydgate.
563. *Hibiscus* = *H. kahilii* Forbes, an upland species on Kauai.
579. *Byronia* = *Ilex anomala* Hook. & Arn. Common on Kauai. When there was an upland forest on Niihau, this could very well have grown there.
611. *Antidesma* = *A. platyphyllum* var. *hillebrandii* Pax & K. Hoffm. Common on Kauai.
617. *Acronychia retusa* = *Pelea anisata* H. Mann. var. *anisata*. Common on Kauai.

William T. Brigham visited Niihau in 1864 and collected *Brighamia insignis*, and doubtless other plants. He sent his specimens to Horace Mann Jr., who combined them with his own collections and arranged and numbered then all as Mann and Brigham collections. At this time, the exploitation of the island as a sheep and cattle ranch had just begun.

William Hillebrand explored widely in the Hawaiian Islands, but he did not visit Niihau. He wrote the basic flora of Hawaii, and noted (1888: 416), "the vegetation...of Niihau is very little known".

Joseph F. Rock did not visit Niihau, but wrote in his *Indigenous Trees of the Hawaiian Islands* (1913, 1973: 87), "The island of Niihau is in a similar state, though not as eroded as Kahoolawe. The vegetation of this small island has, however, disappeared. *Acacia farnesiana* and *Prosopis juliflora* (*Kiawe* or Algaroba) has been planted in the lowlands."

During January 1912 John F.G. Stokes, anthropologist of the Bishop Museum, was allowed to visit Niihau, and while there he made a considerable collection of plants for his brother-in-law, Charles N. Forbes, who published on it in 1913. This Stokes collection, which is in the Bishop Museum, totaled 111 species, including 25 species endemic to the Hawaiian Islands, 39 species indigenous to the islands, 10 species of aboriginal introduction and 37 introduced and naturalized ones.

Harold T. Stearns of the U.S. Geological Survey, visited Niihau from May 23rd to 30th, 1945, to make a geologic and hydrologic survey. He listed (1947: 6) eight trees used in reforestation, and one native, *Erythrina sandwicensis*.

With a biological team from the University of Hawaii, the present writer visited the island from August 9th to 16th, 1947, and alone from March 29th to April 4th, 1949. On both occasions, I had as my guide the knowledgeable and helpful Kalani Niau. Each day for ten or twelve hours we explored on horseback. The trip in August was in the middle of a long drought, and only a few plants were in flower. The second trip was planned to follow a winter rain. I chose to go 6 weeks after the rain, but a three week interval would have been better, as in the six weeks many species had produced shoots, flowers, and were already in fruit.

Kalani Niau, guide and botanical informant to author, 1947.
(*photo by H. St. John*)

Later collections in 1976 and 1977 have been made by John Fay and Charles Christensen.

Vegetation

The original vegetation of Niihau has been decimated by the grazing of goats, sheep and cattle. Native plants of the sandy or salty shores of the island still occur, and perhaps only one (*Scaevola coriacea*) of this group has been exterminated. Native plants of the lowlands, and of the uplands, have been so much depleted that most of the surviving ones can only be found in crevices, or rock faces, or other places, that cannot be reached by grazing animals.

The journals of the early voyagers give only some hints as to the vegetation. From Captain Cook's voyage, March 14, 1779, David Samwell(1967: 1,231) wrote, "Neehaw for the most part consists of low land entirely bare of trees, the Soil is rich and capable of producing all kinds of fruit was it properly cultivated,...We procured yams enough here to serve the 2 ships for bread about six weeks which falls short of the Supply we expected; the Natives cultivate more sweet potatoes than Yams."

Captain Cook himself recorded for February 1, 1778 that the rising ground, "was in a state of nature, very stony, and the soil seemed poor; It was, however, covered with shrubs and plants, some of which sent forth the most fragrant smell I had anywhere met with in this sea".

Handy(1972: 434) says, "On the whole Niihau was a barren, inhospitable island, with little fresh water and no forests on the uplands to supply many of the needs of its inhabitants".

The lowlands still have some trees of *Erythrina sandwicensis, Hibiscus tiliaceus* and *Thespesia populnea*.

The uplands of the mountain mass, rising abruptly to 1,281 feet in height, still have trees of *Dodonaea eriocarpa* var. and *Myoporum sandwicense*; and in the last century had *Cheirodendron trigynum* var. and *Reynoldsia sandwicensis*.

In 1949 the manager, A. F. Robinson, long a resident on the island, told the writer that the mountain had once been forested, and that his ancestors had cut logs and built a log cabin near the mountain crest. Trees that produced trunks big enough to build a log cabin must have been 25 or more feet in height. Just as the *Cheirodendron* and the *Reynoldsia* had been recently exterminated, so several other tree species may well have been destroyed. It is pointless to try to guess what they may have been.

The present vegetation (as seen in 1947 and 1949) is dominated by *Prosopis pallida* which forms, a dry, open, sparse stand over the lowlands (including scattered *Erythrina sandwicensis* trees), and is also found on the upland except near the crest and on the cliffs. On the uplands the only forest remnant is of two patches of *Dodonaea eriocarpa* var. *obtusior*, now only bush-like in stature, and one small stand of *Myoporum sandwicense*. On the western side of the lowland and on the mountain upper ridge there are areas of dry grassland. At the north end of the mountain ridge there is a dense thicket of the introduced *Leucaena leucocephala*, and it also makes a fringe along the western foot of the mountain.

The prominent volcano Kawaihoa, 548 feet high, which is the southern point of the island, may once have had some forest vegetation. If any of it had survived, it was destroyed in World War II when an American aviator dropped a flare which set the peak on fire and burned off its plant cover.

Drift Seed

There were drift seeds on the beaches, particularly on the northeast shore. The writer made on gathering on April 1, 1949, on the beach one mile north of Poooneone. It totaled 3 nuts of *Aleurites moluccana* and 23 seeds of *Mucuna gigantea*, and one of *Strongylodon ruber* Vogel. The *Mucuna* and the *Strongylodon* seeds were good, but the *Aleurites* nuts were without husks and with the seed cavity empty, or with the seed decayed. Evidently they had floated from Kauai, pushed by the northeast tradewind. None of these species were successful in germinating and establishing themselves on the island.

Crops cultivated by the early Hawaiians

Saccharum officinarum L.—*Ko*, for its sweet juice.
Cyperus laevigatus L.—*Makaloa*, to make mats and clothing.
Cocos nucifera L.—*Niu*, for nuts with "milk" and oily meat, numerous uses of leaves, etc.
Musa x *paradisiaca* L. ssp. *normalis* Ktze.—*Maia*, a few plants for the fruit.
Dioscorea alata—*Uhi*, important for its starchy root.
Cordyline terminalis (L.) Kunth var. *ti* (Schott) J. G. Baker—*Ti*, for its sweet root and its leaves.
Artocarpus altilis (Parkins. ex. Z) Fosb.—*Ulu*, scarce, but its fruits edible.
Hibiscus tiliaceus L.—*Hau*, its bark useful for cordage and its sap for medicine.
Thespesia populnea (L.) Soland.—*Milo*, its wood for bowls and utensils.
Ipomoea batatas (L.) Poir.—*Uala*, its starchy roots were an important food.
Cordia subcordata Lam.—*Kou*, its wood was useful.
Morinda citrifolia L.—*Noni*, important source of dye and medicine.

Catalog of the Flora

Algae

Cyanophyta—Blue-Green Algae

Scytonemataceae

Scytonema

Scytonema hofmanii B. & F.
On trunk of *Prosopis pallida*, Mokouia Valley, *St. John 22,820*. It made a close dark coating on the bark and grew intermingled with *Anacystis montana*, while above both were patches of the lichen *Physcia picta*. Determined by W.J. Newhouse.

Chroococcaceae

Anacystis

Anacystis dimidiata (Kutz.) Dr. & Daily
Green globular masses on wet mud, Kaali Spring, 750 ft. alt., *St. John 22,837*. Collected in a mixture with the following. Determined by Dr. F. Drouet.

Anacystis montana (Lightf.) Dr. & Daily
Green globular masses on wet mud, Kaali Spring, 750 ft. alt., *St. John 22,837A*. Collected in a mixture with the preceding. Determined by Dr. F. Drouet.

Chlorophyta—Green Algae

Protococcaceae

Chlorococcum

Chlorococcum infusionum (Schrank) Menegh.
On bark of live branches of 8 m. *Prosopis pallida* tree, mouth of Mokouia Valley, 20 ft. alt., *St. John 22,824*. This was a common yellowish-green crust on the older branches. Determined by Dr. M.S. Doty.

Characeae

Chara

Chara braunii Gm.
In shallow water of flooded *Prosopis* forest, 1 mile W. of Kawaewae, 100 ft. alt., *St. John 23,625*. Determined by Dr. R.D. Wood.

Fungi

Basidiomycetes

Pucciniaceae

Puccinia

Puccinia heterospora Berk & Curt.
On leaves of *Abutilon incanum*, a subshrub growing in dry red dirt, north of Halulu Lake, 30 ft. alt., *St. John 23,647*. Determined by Dr. D.P. Rogers.

Lichens

The lichens were determined by Dr. M.L. Lohman

Lecanoraceae

Lecanora

Lecanora campestris (Schaer.) Hue
On branches of *Prosopis pallida*, Mokouia Valley, 50 ft. alt., *St. John 22,818*. Moist disk pale red-brown, irregularly convex and margin smooth even though subcrenulate when dry. Ascus 45 x 12-14 microns, the wall deep blue. Ascospores 10-14 x 7-9 microns, the wall thick but I⁻. Gonidia 10 microns diam., yellow with KOH. Thallus KOH^+, yellowish, ashy-white when free from bark.

Parmeliaceae

Parmelia

Parmelia conspersa var. *isidiata* (Anzi) Stiz.
On basalt, sea cliff, ½ miles N. of Pueo Point, 1,200 ft. alt., *St. John 23,803A*. Algal layer KOH^+, yellow, strong. Medullary layer KOH^+, yellow in about half of the hyphae and soon bright orange-red; the remaining hyphae unchanging but clear. Brown layer KOH^-.

Parmelia cristifera Tayl.
On basalt, sea cliff, ½ mile N. of Pueo Point, 1,200 ft. alt., *St. John 22,801*. Large, sterile. Medulla KOH^+, yellowish but reaction very slow. Thallus section 400 microns thick, with thick, straight, dark rhisoidal columns in scattered fashion, but none marginal.

Usneaceae

Ramalina

Ramalina faurieana Zahlbr.
On branches of *Prosopis pallida*, Mokouia Valley, 50 ft. alt., *St. John 22,819*. Cortex KOH^+, yellow. Medulla KOH^+, orange red.

Some subcavernous and occasionally perforated as in *R. inflata*. Thalloid exciple KOH^+ yellow. Thecium 45-55 microns. Hymenium 75 microns. Tip of ascus I^+ blue. Paraphasis tip I^-. Ascospores slightly curved, unconstricted, 10-12 x 4-5 microns. In the smaller thalli resemblance is to Nihoa Island, *Caum 75*. The same could be from shrubs. Nothing similar in Magnusson's *Lichens of Nihoa and Necker Islands* (1942), but he treats only material collected on rocks.

Ramalina furcellata (Mont.) Zahlbr.
  On basalt, sea cliff, ½ mile N. of Pueo Point, 1,200 ft. alt., *St. John 22,802a*. Intermingled with *Teloschtes flavidans* f. *glaber*. Scant but few apothecia. Cortex KOH^+ orange-red. Ascospores ellipsoidal, 10 x 4 microns in one, 10-15 x 4 in another, 1.5 mm diameter. Hymenium 45-60 microns deep. Ascus tip I^+ blue. Hypthecium dark.

Ramalina similis Magn.
  On basalt, sea cliff, ½ mile N. of Pueo Point, 1.200 ft. alt., *St. John 22,803*. Clusters larger than given by Magnusson (1945). Entire thallus macroscopically KOH^-, except for soredial emergencies which are KOH^+ red. Cortex 12-18 microns, KOH^-. Medulla, subhymenium, and tips of paraphyses KOH^+ yellow. Hymenium 60 microns. Ascus wall and tips of paraphyses I^+ blue. Ascospores hyaline, ellipsoidal, slightly curved to straight, 10-14 x 4 microns, KOH^- and I^- for both wall and content.

Physciaceae

Physcia

Physcia alba (Fée) Muell. Arg.
  On branches of *Prosopis pallida*, Mokouia Valley, 50 ft. alt., *St. John 22,817*. Ascospores 18 x 8-9 microns. Hypothecium hyaline, KOH^+ yellow. Hymenium I^+ blue-green. Algal layer regular above, irregular below. Thallus as sectioned 170 microns.

Physcia crispa (Pers.) Nyl.
  On trunk of *Prosopis pallida*, Mokouia Valley, 100 ft. alt., *St. John 22,822*. Sterile, pinkish, pale below. Soredia at first marginal, then diffused and thallus at last warty and chinky in center and crisp on

margin. Gonidia 12-14 microns diameter, the layer 60-75 microns. Upper cortex 15-20 microns, the lower very indistinct. Thallus KOH^+ yellow, strong mist. Upper cortex KOH^+ weak. Algal layer $KOH^\pm$ yellow, rhizinae yellowish with brownish strands on treatment.

Physicia pandani Magn.

On branches of *Prosopis pallida*, Mokouia Valley, 50-100 ft. alt., *St. John 22,817a; 22,825; 22,826.* For no. *22,825* which is fertile, thallus sect. 105-140 microns thick. Upper cortex KOH^-. Algal layer KOH^+ yellow with mist. Medulla KOH^-. Weak lower cortex KOH^+ rusty violaceous.

Physcia picta (Sw.) Nyl.

On trunk of *Prosopis pallida*, Mokouia Valley, 100 ft. alt., *St. John 22,820; 22,821; 22,823.* For no. *22,821* which is fertile, thecium 70-80 microns, deep blue with I. Ascospores fusoid, 2-celled, 18 x 5-7 microns. Thallus KOH^+ yellow. Soredia KOH^+ yellow. Cortex KOH^+ reddish-brown above and below. Gonidial layer KOH^{++} yellowish-green. Medulla KOH^-.

Teloschistaceae

Telochistes

Teloschistes flavicans f. *glaber* Vain.

On basalt, sea cliff, ½ mile N. of Pueo Point, 1,200 ft. alt., *St. John 22,802.* Sections averaging 0.6 x 0.2 mm. with algae layer irregular and gonidia on large hyphae in lax medulla. Cortex lacking in spots. Cortex and hyphae of gonidia in lax area KOH^+ red (with crystals). In water mount cortex yellow. All portions Pb^-.

Mosses

Pottiaceae

Weisia

Weisia viridula Hedw.
On basalt sea cliff, ½ mile N. of Pueo Point, 1,200 ft. alt., *St. John 22,800*. Determined by H.A. Miller.

Hypnaceae

Ectropothecium

Ectropothecium viridifolium Bartr.
By small spring on wet basalt ledge with *Adiantum*, Mokouia Valley, 300 ft. alt., *St. John 22,843*. Determined by E.F. Bartram.

Ferns and Fern Allies

(A key to the recorded genera was provided, but data concerning collection sites, etc. was unavailable except for *Marsilea*.)

Key to the Genera

Fronds like clover leaves. *Marsilea*
Fronds not so.
   Stems triangular, slender, forking, lacking an expanded frond.
                                                               Psilotum
   Stems not so; frond expanded into leaves.
      Fruit dots (sori) of slender lines on back of frond.
         Fronds white-floury beneath. *Pityrogramma*
         Fronds green beneath. *Doryopteris*
      Fruit dots (sori) broader.
         Fruit dots (sori) oblong, on margins of frond. *Adiantum*
         Fruit dots (sori) on back of frond, round.
            Veins of frond with free tips. *Nephrolepis*
            Veins of frond forming a net. *Dryopteris*

Adiantum

Adiantum sp.
‘Iwa ‘iwa

Doryopteris

Doryopteris decipiens (Hook.) J. Sm.
Kalamoho, alamoho

Dryopteris

Dryopteris dentata (Forsk.) C. Chr.
(*Pai‘i‘iha* elsewhere)

Marsilea

Marsilea villosa Kaulf.
‘Ihi la‘au (=sorrel, medicinal)
Dried muddy border of Loe Lake, 2 mile N. of Puuwai, 10 ft. alt., *St. John 23,600*.

Nephrolepis

Nephrolepis exaltata (L.) Schott
Ni‘ani‘au, palapalai, (*kupukupu* elsewhere)

Psilotum

Psilotum nudum (L.) Beauv.
Moa.

Pityrogramma

Pityrogramma calomelanos (L.) Link
Palapalai

Conifers

Gymnospermae

Araucariaceae (Araucaria Family)

Araucaria

Araucaria columnaris (Forst. f.) Hook.
Determination tentative, based on field observation. Recorded by C.S. Judd (1932) as Norfolk Island Pine. Observed in 1947 by the present writer, but not collected.

Flowering Plants

Angiospermae

Monocotyledones

Pandanaceae (Screw Pine Family)

Pandanus

Pandanus odoratissimus var. *levigatus* Martelli
Hala
This species was observed in 1912 by Stokes who saw one tree. The present author noted one clump of trees on sands near lake, ½ mile N.E. of Kiekie, 20 ft. alt., *St. John 22,790*. Other clumps, remembered by the inhabitants, had died out in dry spells.

Potamogetonaceae (Pondweed Family)

Potamogeton

Potamogeton pectinatus L.
Limu-alolo (=the limp kelp)

In temporary flood water between flooded *Prosopis* trees, Loe Lake, 2 miles N. of Puuwai, 10 ft. alt., *St. John 23,597*. First record in the Hawaiian Islands. Perhaps brought to the islands by migrating water fowl.

Ruppiaceae

Ruppia

Ruppia maritima var. *pacifica* f. *pacifica* St. John & Fosberg
   Limu-pa-kai (=the kelp of the sea)
   Rooting, submerged in 6 inches of water, salty inlet, Leahi, *St. John 23,620*.

Gramineae (Poaceae) (Grass Family)

Key to the Genera

Fruits rounded, covered with barbed bristles. *Cenchrus*
Fruits not so.
  Tree-like, with hard woody stalks (bamboo). *Schizostachyum*
  Not tree-like.
    Stems solid, with sweet pith. *Saccharum*
    Stems hollow.
      Foliage sticky, hairy. *Melinis*
      Foliage not so.
        Spikelets falling free from the two lower scales.
          Spikelets attached directly to one side of finger-like spikes.
            Spikelets with sterile florets above the fertile one. *Chloris*
            Spikelets with only a single fertile floret. *Cynodon*
          Spikelets short stalked.
            Spikelets one-flowered. *Sporobolus*
            Spikelets several-flowered *Eragrostis*
        Spikelets falling with all their scales.
          Outer two scales of spikelet membranous.
            Spikelets surrounded by bristles. *Setaria*
            Spikelets not so.

Three lower scales of spikelet bristle-tipped.
Echinochloa
Lower scales of spikelet not so.
Spikelets long-silky, hairy.
Rhynchelytrum
Spikelets not so.
Base of spikelet swollen and knob-like.
Eriochlora
Base of spikelet not swollen and knob-like.
Spikelets in two rows on one side of the axis. *Digitaria*
Spikelets not so.
Spikelets in finger-like spikes.
Paspalidium
Spikelets in open clusters.
Panicum
Outer two scales of spikelet rigid.
Base of spikelet sharp, and the bristles hairy.
Heteropogon
Base of spikelet not so.
Of the lowest pair of spikelets, one fertile and stalkless, the other male and stalked.
Andropogon
The lowest pair of spikelets alike, male or neuter. *Cymbopogon*

Andropogon

Spikes numerous, forming a dense silky head. *A. barbinodis*
Spikes 2 or more, finger-like.
    Outer scale of floret deeply pitted. *A. pertusus*
    Outer scale of floret not deeply pitted, spikes long, finger-like, silky. *A. sericeus*

Andropogon barbinodis Lag.
    Mauʻu-haole
    In tufts on red dirt, inflorescence silvery, north of Halulu Lake, 30 ft. alt., *St. John 23,646*. A pasture grass, introduced from Mexico.

Andropogon pertusus (L.) Willd.
   Paniau, 1,100 ft. alt. *St. John 23,583*; Kiekie, 50 ft. alt., *St., John 22,845*. An introduced grass, cultivated on pastures as forage. It has no Hawaiian name.

Andropogon sericeus R. Br.
   Huelo-popoki (=tail of the cat)
   Tufts on red dirt flats, very blue in color, north of Halulu Lake, 30 ft. alt., *St. John 23,643*. Forage grass, introduced from Australia.

Cenchrus

Burs 8-12 mm. long; leaves 4-12 mm. wide; perennials.
                                                                C. agrimonioides
Burs 4-7 mm. long; leaves 3-9 mm. wide; annuals.   *C. echinatus*

Cenchrus agrimonioides Trin.
   Pennisetum calyculatum Spreng.
   Agrimony sandbur.
   Collected in 1826 by Lay and Collie, botanists of the *H.M.S. Blossom*, under Capt. Beechey and recorded (1832: 101) under the latter botanical name as "Found in Oneeheouw". It has not been noted recently and is probably extinct.

Cenchrus echinatus L.
   ʻUmeʻalu
   Kaali, *Stokes*; a single bur mixed with *Hydrocotyle*, the collection determined by and recorded by St. John (1931: 3). Abundant and pestiferous on lowland flats, especially in sandy areas. Observed by St. John at Kalanihale Kaununui, Paniau, Poooneone and Kamalino. These lie on the low, coastal areas of three sides of the island, except for Paniau where this bur-fruited grass was noted at the hill summit at the brink of the precipice on the windward side.

Chloris

Third scale of spikelet broad and inflated at tip.　　*C. inflata*
Third scale not as above.
　Stems up to 50 cm. tall; blades 1-2 mm. wide.　　*C. divaricata*
　Stems 1-1.5 m. tall; blades 3-5 mm. wide.　　*C. gayana*

Chloris divaricata R. Br.
　Common in pastures, introduced, 1 mile W. of Kawaewae, 50 ft. alt., *St. John 23,623*. This collection differed materially from the grass which is a common weed in the lawns on Oahu, one which was previously first recorded and identified as *C. divaricata* by Hosaka (1936: 126). The herbarium and library resources in Honolulu were insufficient to enable the writer to identify these adventive grasses with surety. Mr. C.E. Hubbard of Kew kindly identified the specimens sent to him. He determined *St. John 23,623* as *C. divarivata* R. Br.; stating that it was matched with the type collection from Australia, *Brown 6,244*; and *St. John 24,832* (location not given) as *C. cynodontoides* Balansa in Bull. Soc. Bot. France 19: 318, 1872, described from New Caledonia, and now also known from Fiji. Since these specimens have been confused, the writer offers a key to enable students to distinguish them.
Lemma margin glabrous; first glume 1 mm. long; second glume 2-2.5 mm. long; spikes 3-5, flexuous or sinuous, 9-15 mm. long.
　　　　　　　　　　　　　　　　　　　　　　　C. divaricata
Lemma margin pilose-ciliate; first glume 1.5 mm. long; second glume 2.5-3.5 mm. long; spikes 5-9, stiff, nearly straight, 5-10 cm. long.　　　　　　　　　　　　　　　　　　*C. cynodontoides*

　There are in the Bishop Museum two other collections of *C. divaricata* from the Hawaiian Islands, both from Lanai and collected by G.C. Munro, one of these dated 1924.

Chloris gayana Kunth
　Rhodes grass.
　Cultivated in pastures, Kiekie, 50 ft. alt., *St. John 23,659*. Observed, but not collected, at Apana Valley. An introduced grass, from Africa, used for forage.

Chloris inflata Link
Weed in grassy flat, Kii, 20 ft. alt., *St. John 23,564*. Also observed at Halulu Lake, Halalii Lake and Poooneone. An introduced weed, common, but drying up and disappearing in the dry season.

Cymbopogon

Cymbopogon refractus (R. Br.) A. Camus
Manienie mahiki(=grass, projecting)
Pasture grass, ½ mile northeast of Kiekie, 400 ft. alt., *St. John 22,794*. An Australian grass, introduced and cultivated for forage.

Cynodon

Cynodon dactylon (L.) Pers.
Capriola dactylon (L.) Ktze.
Manienie
Recorded by Forbes (1913: 20) from Stoke's trip, but apparently no specimen was preserved. Observed, but not collected, by St. John, at Kalanihale, Paniau, Kaununui, Poooneone and Kamalino. It is well established and common from the flats near the beach to the tops of the hills.

Digitaria

Digitaria setigera Roth
Recorded by Forbes (1913: 20), from Kaali, the specimen is now verified. It was not observed by the writer on his visit during a dry season. A second collection by Stokes, from foot of mountain on west side was determined by Forbes as *Panicum pruriens* Trin., but the present writer redetermined it as *Panicum niihauense* St. John.

Echinochloa

Echinochloa colona (L.) Link
Mau'u alolo
Muddy spot in basalt talus, steep rocky shore E. of Kalaalaau Valley, 75 ft. alt., *St. John 22,746*. A grass introduced from tropical America.

Eragrostis

Spikelets 1.5-2mm. long; inflorescence loose.   *E. tenella*
Spikelets longer.
   Spikelets 3-10mm. long; inflorescence narrow, compact; leaves 1-2 mm. wide.   *E. niihauensis*
   Spikelets 5-10mm. long; inflorescence narrow or loose; leaves 5-10mm. wide.   *E. variabilis*

Eragrostis niihauensis Whitney in Occas. Pap. Bishop Mus. 12(5): 6. fig. 2. 1936.
Kawelu (= ragged)
Endemic to Niihau. Type specimens were collected by Kalua Keale in February 1912 (represented by two sheets in the Bishop Museum). The present writer's specimens were collected at Kawaihoa Point. In one collection the plants were common, tufted and on a steep, grassy, west slope at an altitude of 100 ft., *St. John 22,783*. The other was collected at 200 ft. alt., and the plants were occasional in a *Heteropogon* grassy slope between 200-548 ft. alt. The tufts consisted of 20-40 culms, with the old leaves marcescent and the blades involuted and rigid, *St. John 23,616*.

Eragrostis tenella (L.) Beauv. ex Roemer & Schultes
    E. amabilis (L.) Wight & Arn. ex. Hook. & Arn. 1838.
Sandy flat near shore, Kii, 10 ft. alt., *St. John, 23,663*; in grassy vales, Kaaliwai, 30 ft. alt., *St. John 23,569*. Also observed at Kamalino and Leahi. An introduced annual grass with thread-like stems. It is introduced from the tropics of the Old World.

Eragrostis variabilis (Gaud.) Steud.
In crevices of basalt cliff, Kaalipuaa, or in tufts of 10-15 stems, 750 ft. alt., leaves inrolled, bright green, *St. John 23,577*; also Niihau, in 1912, *Kalua Keale* (from Forbes), cited by Hitchcock, Mem. Bishop Mus. 8(3): 137, 1922, but not now in the Bishop Museum.

Eriochloa

Eriochloa procera (Retz.) C.E. Hubb.
In dried muddy bed of old Makanikahau Reservoir, Apana, 400 ft. alt., *St. John 23,634*. A forage grass, introduced from tropical Asia.

Heteropogon

Heteropogon contortus (L.) Beauv. ex Roemer & Schultes
Pili
Ledges on sunny basalt cliff, Kaalipuaa, 700 ft. alt., *St. John 23,578*; top of cliff, in basalt crevice, occasional, Kawaewae, 290 ft. alt., *St. John 22,738*. Observed by St. John at Kalanihale, Kaali Spring, Mokouia Valley and Kawaihoa Point. It is excellent forage, but has been much reduced by grazing. Good specimens were found only in somewhat inaccessible crevices. Originally it was probably abundant in the drier lowlands.

Melinis

Melinis minutiflora Beauv.
At times this has been cultivated for forage at Kiekie. A.F. Robinson informed the writer that it did well in good seasons, but always died out when a drought came.

Panicum

Perennials.
With running root stocks; leaves 10-35 mm. wide; plants 1-2 m. tall.
P. maximum
Tufted; leaves 7-9 mm. wide; plants 50-53 cm. tall.
P. niihauense

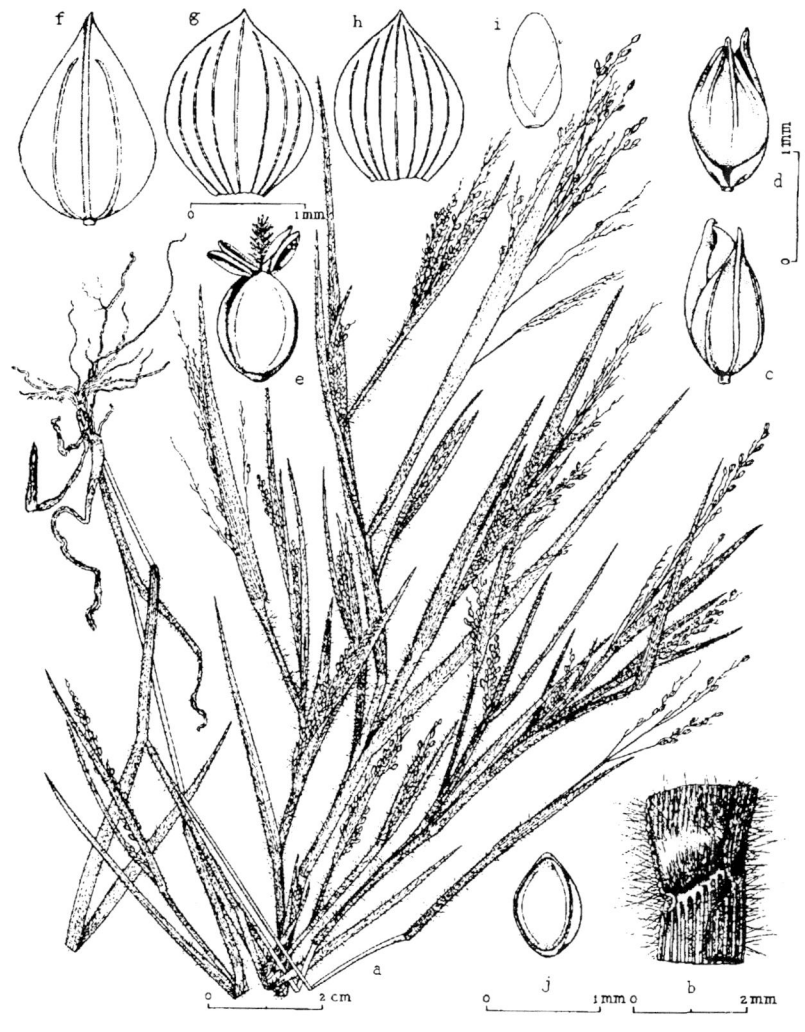

Panicum heupueo, from the holotype: a, habit X 1: b, blade and sheath X 10; c, spikelet X 20; d, spikelet X 20; e, floret X 20; f, first glume, dorsal view, X 20; g, second glume, dorsal view, X 20; h, sterile lemma, dorsal view, X 20; i, sterile palea, ventral view, X 20.

Annuals, or short lived.
   Spikelets not hairy, 1.2-1.5 mm.. long; leaves 2-4 mm. wide.
P. heupueo
   Spikelets hairy.
     Spikelets with only a few long hairs; 1.5 mm. long; leaves 2-5 mm. wide.   *P. colliei*
     Spikelets densely long, hairy, 3-4 mm. long; leaves 2-4 mm. or 8-10 mm. wide.
        Stems and leaves with soft, long hairs; 3 mm. long; leaves 8-10 mm. wide.   *P. torridum*
        Stems without or with only short hairs; 4 mm. long; leaves 2-4 mm. wide.   *P. ramosius*

Panicum colliei Endl.
   Kaiʻoiʻo (= growing irregularly)
   Rocky basalt knoll, one-half mile W. of Kii, 50 ft. alt., *St. John 23,565*. Between basalt ledges on rocky knoll, Kii, 100 ft. alt., *St. John 23,668*, and *23,666A*.

Panicum heupueo St. John, Pacif. Sci. 13: 156-158, *fig. 1.* 1959.
   Heʻupueo (= fine hairs on the owl); *hakonakona*, perhaps in error for *P. torridum* Gaud.
   A recently described endemic for Niihau. The type specimen is from Kii, between basalt rocks, on a rocky knoll, 100 ft. alt., April 2, 1949, *St. John 23,666*. Also from the same locality and time, *St. John 23,665*, and *23,670*.

Panicum maximum Jacq.
   Mauʻu pulumi
   Planted ½ mile W. of Kii, two meter clumps by trail, 110 ft. alt., *St. John 23,605*. Here a forage grass introduced from Africa.

Panicum niihauense St. John, Occas. Pap. Bishop Mus. 9(14): 5-6. pl.1. 1931.
   Lau ehu (=leaf, reddish or sand-colored)
   Recorded by Forbes (1913: 20) as *P. nephelophilum* Gaud var. *xerophilum*?, but this was a misdetermination of Hillebrand's variety. The type and only original collection of *P. niihauense* was "Foot of

A view of the east coast pali of Niihau.

Lobelia niihauensis
(photo taken on Kauai)

Cordia sebestena

Ipomoea batatas

Capparis sandwicensis

Eragrostis variabilis

The middle savannah of Niihau with Kauai in the background.

The southern end of Niihau.

Heliotropum anomalum

Scaevola on a beach on Niihau.

The northeast coast of Niihau.

The upper west coast and a seasonal lake of Niihau.

Mt. on W. Side". Another specimen with the same data was misdetermined by Forbes as *Panicum pruriens* Trin. Recently the species was rediscovered ½ mile west of Kii, only on rocky basalt knoll, in clumps of 20-30 stems, with *Sida cordifolia* and *Ipomoea*, 50 ft. alt., *St. John 23,566*. The numerous arching stems and leaves are pinkish and hoary pubescent. They crown the low rocky knolls, overtop all other plants and are beautiful. Also it was found, more sparingly, southwest of Kii, on coral sand dunes, where it sprawls on the sand. Here the leaves were at first reddish, but later blue-green. This collection was at 50 ft. alt., *St. John 23,672*.

Panicum ramosius Hitchc.

The type was collected by Lay and Collie on the Voyage of the *Blossom*. Though published as simply from the Sandwich Is., it was cited by Hillebrand, who had studied at Kew, as "Niihau (Lay and Collie)". There are no recent collections from Niihau, where it is probably extinct, but its range has now been extended to include Kauai, Oahu, Molokai and Lanai.

Panicum torridum Gaud.

Hakonakona (= coarse); also *Kai oʻiʻo* (= thick growing; though *kai* = sea; and *oʻiʻo* = a fish, eaten raw, mashed, salted, then mixed with water to make a thick paste)

On basalt cliff, in wet mud, below spring, Kaalipuaa, 750 ft. alt., *St. John 23,576*; between basalt rocks on rocky knoll, Kii, 75 ft. alt., *St. John 23,667*, and *23,669*.

Paspalidium

Paspalidium radiatum Vickery
Panicum radiatius St. John
Cultivated pasture grass, Kiekie, 50 ft. alt., *St. John 23,661*.

Rhynchelytrum

Rhynchelytrum repens (Willd.) Hubb.
Tricholaena repens (Willd.) Hitchc.
Puaʻulaʻula (= flower, red)

On basalt rocky knoll, below Kaala Cliff, 50 ft. alt., *St. John 23,579*. Also observed at Paniau. The flowering tops of this grass are brilliant with reddish hairs. It is a good forage grass after a wet spell, but quickly dries up and vanishes. The species has been introduced from South Africa.

Saccharum

Saccharum officinarum L.
 Ko

Not collected, but observed, cultivated in gardens at the village, Puuwai. These patches were of sugar cane of modern commercial varieties, recently introduced from Kauai. It was learned that in the old days the natives cultivated two varieties, known as *ko kea*, (white sugar cane), and *ko ula*, (red sugar cane). These old Hawaiian varieties have now become extinct on the island.

Samwell, who visited Niihau with Capt. Cook in 1778, recorded seeing "a few plantations of sugar cane". Forbes reported (1913: 20) that, "Dr. Brigham remembers seeing the native sugar cane growing in coral caves on the lowlands".

Capt. Portlock recorded (1789: 84), "After examining the wells, I made an excursion into the country, accompanied by Abbenooe, and a few of the natives. The island appears well cultivated; its principal produce is yams. There are besides, sweet potatoes, sugar cane, and the sweet root called tee by the natives."

In 1949, Mrs. Mary K. Pukui told the writer of accounts of Niihau she had heard from her grandmother, Poai, who had accompanied Queen Emma on a visit to the island in about the year 1860. Later, Poai often reminisced of a kind of sugar cane seen there, called "*ko eli o Halalii*". It grew in flats near the shore where the wind blown sand would cover it up to the tips of the leaves. When wanted, the people would find the green tips, then excavate the sugar cane stalks from the dune sands.

Schizostachyum

Schizostachyum glaucifolium (Rupr.) Munro
 'Ohe
 Observed but not collected at Puuwai. This species of many Pacific islands is of old cultivation by the natives of Polynesia.

Setaria

Setaria verticillata (L.) Beauv.
 Chaetochloa verticillata (L.) Scribn.
 Doubtless this pestiferous grass has a native Niihauan name, but it was not ascertained. It was collected by *Stokes* at Kaali and by the present writer at the wet seep by Kaali Spring, 750 ft. alt., *St. John 22,835*. It was also observed at Mokouia Valley and Apana Valley in the uplands and at Kalaalaau Valley, Puuwai, Nonopapa and Leahi in the lowlands. It is a common pest, with barbed fruits that adhere to man or animal. It was introduced from Europe.

Sporobolus

Sporobolus virginicus (L.) Kunth
 Mahiki
 In stabilized dunes above the beach, Kalanihale, ⅛ mile S. of Puukole Point, 15 ft. alt., *St. John 22,710*. Also observed at Poooneone, Nonopapa, Kamalino and Kawaihoa Point. It is common at the top of the sandy beaches. It is indigenous and is a plant of pantropic distribution.

Cyperaceae (Sedge Family)

Key to the Genera

Scales of flowering spikelet in two ranks.   *Cyperus*
Scales of flowering spikelet in a spiral.
 Spikelets single, terminal.   *Eleocharis*

Spikelets several.
  Nutlets without bristles.   *Fimbristylis*
  Nutlets with bristles at base.   *Scirpus*

Cyperus

Styles 2; nutlets with 2 convex faces.
  Cluster of spikelets on one side; stem leafless.   *C. laevigatus*
  Cluster of spikelets terminal; near them 3-6 leaves.
                                                  C. polystachyus
Styles 3; nutlets with 3 angles and faces.
  Spikelets long, slender, in loose clusters.   *C. rotundus*
  Spikelets lance-shaped, in dense heads or loose spikes.
    Spikelets in short dense heads.
      Spikelets 0.6 mm. wide, in cylindric heads.   *C. phleoides*
      Spikelets 4-5 mm. wide, in rounded heads.
                                               C. trachysanthos
    Spikelets in loose spikes.   *C. javanicus*

Cyperus javanicus Houtt.
    C. caricifolius Hook. & Arn.
    C. pennatus Lam.
    'Ahu'awa, 'ehu'awa

Kaali, *Stokes*; rocky edge of dried salty inlet, Nonopapa, 10 ft. alt., *St. John 22,762*. Also observed by the border of an exsiccated mud flat, opposite the mouth of Kapaka Valley, northeast of Mokouia. The type collection of *C. caricifolius* Hook. & Arn. was obtained by Lay and Collie on either Oahu or Niihau. In the description it was specified only as from the Sandwich Islands. The type, which is at Kew, has not been studied, but the interpretation of Scanlan (1942: 38) in reducing it to the synonymy of *C. javanicus* is here followed.

Cyperus laevigatus L.
    C. mucronatus Vahl.
    Makaloa

Kaali, *Stokes*; two sheets, one with a hank of coiled culms many decimeters long; in dry, subalkaline lake bed, north shore of Halalii Lake, rare, *St. John 22,749*. This species was formerly abundant along the shores of the several temporary shallow lakes and ponds.

The soft, pliant culms were the basis of a plaiting art carried out by the Hawaiian natives with those of Niihau long excelling in the making of fine mats, much prized by the Hawaiian chiefs, though mats of *makaloa* were occasionally made on the other islands. The mats were a natural straw color and were woven in a variety of weaving patterns. Since *wauke* (*Broussonetia papyrifera*) was said not to grow on Niihau, the *makaloa* mats were used for clothing.

This native swamp plant, the *makaloa*, was to a certain degree cultivated, as Stokes found (Forbes, 1913: 19) that, "The areas of *Cyperus laevigatus* which used to be tended with some care are being crowded out by another species, as well as by sheep, except where Mr. Robinson has protected it by fencing as of historical interest". The writer was informed that formerly open barrels or cylinders were placed over plants in order to force then to grow extra long stems. In 1947 the writer found the protective fence around a large patch on Halulu Lake broken and largely gone. The numerous sheep have now almost exterminated the species. My guide Kalanipio Niau hunted in several of the old patches, but not a culm was found. Finally on the shore of Halalii Lake on the border where drooping branches of *Prosopis pallida* made a thorny thicket so dense that even hungry sheep were excluded, a few tiny plants were found.

On the other islands the plant is called *ehuawa* and only the woven mats are called *makaloa*. Yet on Niihau both the plant and the mats are called *makaloa*, an appellation obtained and carefully verified. Mr. Aylmer F. Robinson told the writer in 1947 that as late as 1917, at the request of his father, the natives had made some *makaloa* mats, but the product was distinctly inferior to the older ones. None have been made recently.

In 1947 Mrs. Mary K. Pukui told the writer that in early times the natives on Niihau clothed themselves in *makaloa* mats because they had no *wauke*. Both Stokes in 1912, and the writer in 1947 found bushes of *wauke*, *Broussonetia papyrifera*. It may well have been cultivated there in aboriginal times, yet due to the aridity it may have been too scarce to supply the clothing needs of the people.

Cyperus phleoides var. *hawaiiensis* (Mann) Kükenth.

In basalt crevice of steep cliff, Kaaliwai, 750 ft. alt., *St. John 23,570*. No native name was obtained for this insignificant, grass-like plant, 20-25 cm. in height. The combination for the binomial has commonly been attributed to Hillebrand (1888: 46), but its first and valid publication was by Mann (1867: 208).

Cyperus polystachys var. *texensis* (Torr.) Fern.
   C. polystachys of Forbes' list (1913: 20).
  Kaali, *Stokes*; native name was not recorded.

Cyperus rotundus L.
  Pipi wai (*pipi*= beef; *wai*=water)
  Two miles north of Loe Lake, grassy flats, Puuwai, 20 ft, alt., common weed, much sought and dug up by the wild pigs, *St. John 23,601*.

Cyperus trachysanthos Hook. & Arn.
  Ka'a
  Kaali, 300-325 ft. alt., *Stokes*; Loe Lake, two miles north of Puuwai, exsiccated mud of flood plain, 10 ft. alt., herbage glutinous, *St. John 23,599*; Halalii Lake, south end, on dry coral ledge, 50 ft. alt., leaves and culms glutinous at base, *St. John 22,732*.

Eleocharis

Eleocharis calva var. *australis* (Nees) St. John in Pacif. Sci. 13: 159-163. Fig. 2. 1959.
  Kohekohe
  Marshy border of temporary flood water Loe Lake, two mile north of Puuwai, 10 ft. alt., plants up to 12 dm. tall, basal sheaths red, *St. John 23,598*. There have been a few previous collections of this plant in the Hawaiian Islands, and it has generally been thought to be an introduction. The manner of its occurrence in Niihau seems to mark it as a native variety.

  Each stem of this sedge is surrounded by a handsome basal sheath which is a bright red color. Both the Rev. Henry Judd, in 1950, and the writer learned that the natives had used these red cylinders to make ornamental designs in their *makaloa* mats.

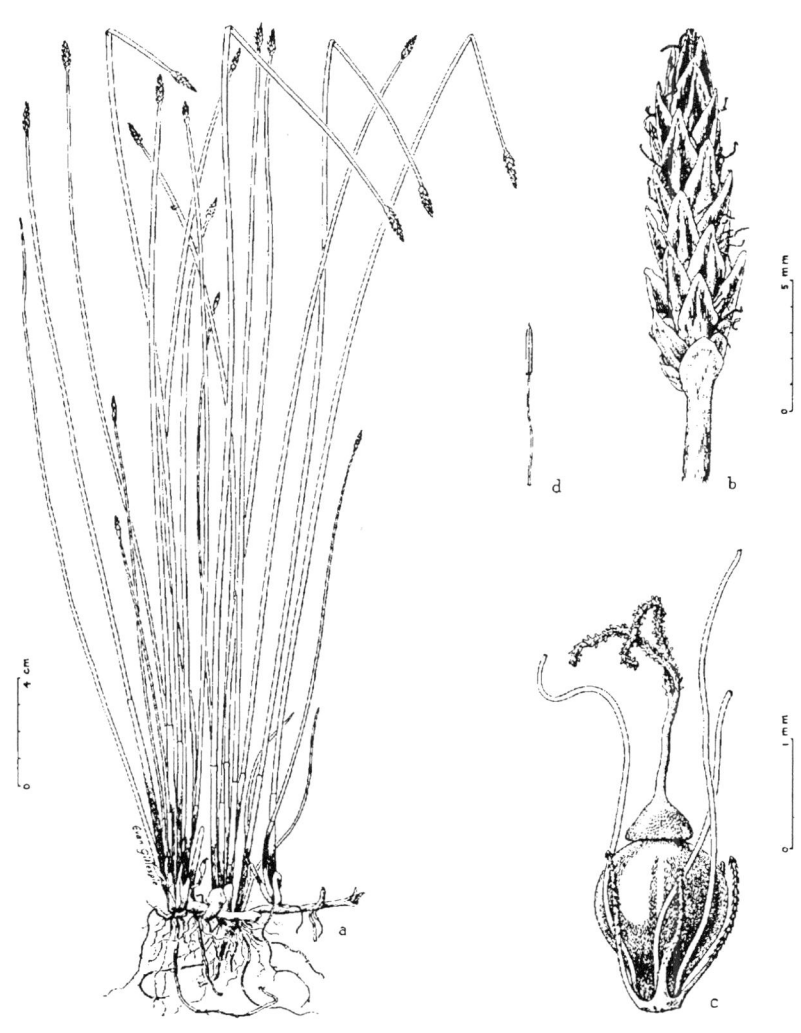

Eleocharis calva var. *australis*, from St. John 23,598: a, habit X 1/2; b, spike X 5; c, achene X 20; d, stamen X 5.

Fimbristylis

Fimbristylis pycnocephala
F. cymosa sensu Hillebrand and Hawaiian botanists, not of R. Br.
F. cymosa var. *pycnocephala* Kükenth. ex Skottsb.
F. umbellato-capitata sensu Mann (1867), non Steud. (1855).
F. cymosa var. *umbellato-capitata* (Steud.) Hbd., sensu Hbd., non Steud.

Mau'u 'aki'aki

Thicket near beach, Poooneone, 20 ft. alt., *St. John 22,744*. First reported by Hooker and Arnott (1832: 98) as collected by Lay and Collie, "both in Oahu and Oneeheow".

As indicated by St. John (1952: 150, fig. 3), *F. cymosa* R. Br., an Australian species described from a collection by Robert Brown from the Gulf of Carpentaria and the Prince Wales Islands, is a very different plant, characterized by trigonous achenes, transversely rugulose; spikelets single or less than half of them paired, etc. Apparently all subsequent botanists have followed Hillebrand in accepting this as the abundant littoral species in Hawaii. He separated the reduced plants of the saltier and more exposed habitats as *F. pycnocephala* Hbd., characterized by having the inflorescence condensed and usually monocephalous. Somewhat further back from the shoreline, and in more protected habitats the plant grew taller and formed looser inflorescences, and these plants Hillebrand called *F. cymosa* R. Br. No fundamental differences between these Hawaiian plants have been detected, and by the shore one can find every transitional state between them. Consequently, they are here treated as one species, the first available name for which is *F. pycnocephala* Hbd. *Fimbristylis umbellato-capitata* Steud., described from Mauritius, in the Indian Ocean, also seems to be a misapplied name when attached to Hawaiian specimens, as was done by Mann, and doubtfully by Hillebrand.

Scirpus

Stems triangular.
    Fruiting spikelets in a naked terminal cluster.   *S. californicus*
    Fruiting spikelets among many small leaves.   *S. paludosus*
Stem round in cross section.   *S. validus*

Scirpus californicus (C.A. Meyer) Steud.
Neke
In clumps on dried mud flat; the first valley east of Mokouia Valley, 20 ft. alt., *St. John 22,842*. Though it is not completely proven, the writer considers this species a recent introduction from America, since it is still infrequent and the first known collection of it was that by Forbes at Pukoo, Molokai in 1912.

Scirpus paludosus A. Nels. var. *digynus* (Hbd.) Beetle
Native name not recorded.
South of Halalii Lake, edge of dry lake bed, 20 ft. alt., *St. John and Tuthill 22,756*.

Scirpus validus Vahl
S. *lacustris* of Forbes' list, not of Linnaeus.
Ponds, southern end, *Stokes*. Doubtless also called by the Hawaiian *neke*.

Palmae (Palm Family)

Cocos

Cocos nucifera L.
Niu
Observed but not collected by Stokes and by the present writer. Coconuts are cultivated in the village of Puuwai, and at a few other spots, such as Halalii Lake. One large grove was found near Keawanui Bay. Dr. H.T. Stearns reported (1947: 6-7), "Small groves of coconuts and date palms thrive in the lowlands, where the water table is close to the surface. They have to be watered for about five years to establish their roots before they will survive a drought." By custom the fresh nuts are not taken by the inhabitants, but are reserved for periods of severe drought, when they are used to supplement the dwindling water supply. The early voyagers do not mention the coconut, so it is probably of recent establishment upon the island.

Phoenix

Phoenix dactylifera L.

Niu kahiki (=foreign coconut)

Observed by H.T. Stearns; see the quotation under *Cocos,* given above. Date palms were observed by the writer at Keawanui Bay, Kiekie and Halalii Lake. A cultivated fruit tree introduced from Africa.

Pritchardia

Pritchardia aylmer-robinsonii St.John in Pacif. Sci. 13: 163-165. Fig. 3. 1959.

Wahane, or occasionally *Hawane*

The type locality is on a knoll protected by basalt boulders, south ridge, Mokouia Valley, 875 ft. alt., *St. John 22,813*. This species was observed on the north slope of Kapaka Valley and a single, old, but damaged, tree 15 meters tall, at 250 ft. alt., was observed in Haao Valley. This is an attractive and interesting palm tree, peculiar to Niihau Island.

Pritchardia remota Becc.

These palm trees, forming a handsome line by the residence, at Kiekie, are known by the Robinson family to have been obtained on the island of Nihoa, of the Hawaiian Leeward Islands. They were observed but not collected.

George C. Munro (1952: 3) relates, concerning Popoia Island, off Kailua, Oahu, "Amongst others I planted...two plants of the Nihoa palms (*Pritchardia remota*), seed of which I had secured on Niihau in 1939. Seed had been brought from Nihoa Island in the late 1800s by Mr. George Gay and planted on Niihau where one had grown into a tall tree..." He recorded that both of the trees planted on Popoia Island died.

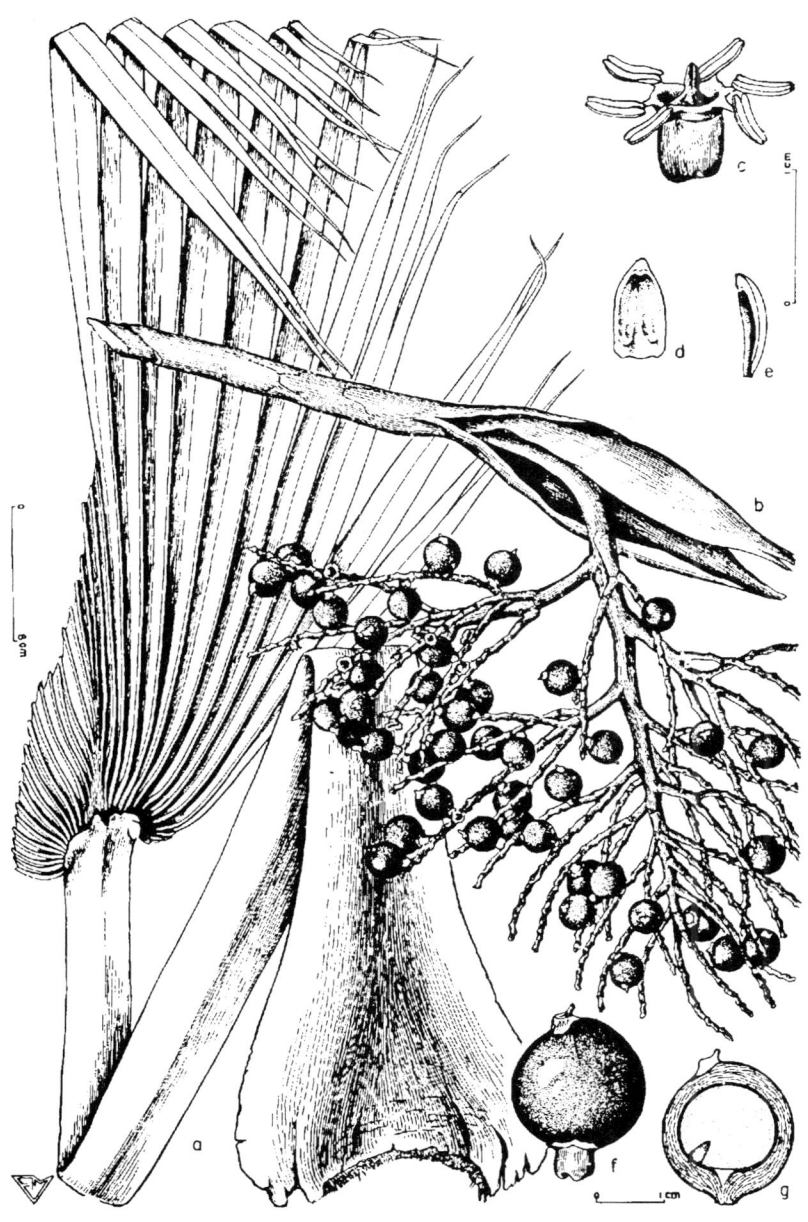

Pritchardia aylmer-robinsonii, from the holotype: *a*, leaf X 1/4;   *b*, fructescence X 1/4; *c*, flower X 2; *d* and *e*, petals X 2; *f*, fruit X 1; *g*, fruit in longtitudinal section X 1.

Araceae (Arum Family)

Colocasia

Colocasia esculenta (L.) Schott var. *antiquorum* (Schott) Hubb. & Rehd.
Taro
Planted in muddy seepage, Kaali Spring, recently introduced from Kauai, 750 ft. alt., *St. John 22,831*. This seems to be of the horticultural variety *Ulaula poni* of Whitney, Bowers and Takahashi (1939: 51).

The fact that taro was not grown in Niihau in aboriginal times is indicated by statements of Capt. Portlock (1789: 86). He presented "Abbenooe", the chief of Niihau, with a "red baize and two large towes" which the chief immediately forwarded to King Ta'aao of Kauai. In return the king sent by canoe hogs, taro, and sugar cane. The present included the only taro seen or received while bartering at Niihau.

Nowadays poi in quantity is regularly transported by boat from Kauai. Niihau has no wet areas for irrigated taro patches, so the several individual plants now grown in the seepage from Kaali Spring are more specimen plants or occasional delicacies than crop plants to furnish food supply.

Bromeliaceae (Pineapple Family)

Ananas

Ananas comosus (Stickm.) Merr.
Hala
Not collected, but observed in cultivation in the vegetable gardens of the natives at the village, Puuwai. The variety was the "Smooth Cayenne" of recent introduction to Niihau, and not the older one first obtained by the Hawaiians. It is native to South America.

Liliaceae (Lily Family)

Cordyline

Cordyline terminalis (L.) Kunth var. *ti* (Schott) J.G. Baker
  La'i, that is the plant, but the root is called *ki*
  One mile N.E. of Leahi, 70 ft. alt., with *Artocarpus* in sinkhole in dry limestone flat, *St. John 22,778*. Probably a relict from the old Hawaiian culture. Breadfruit, sugar cane, and now *ki* are recorded as growing, cultivated in these moist sinkholes, the only moist spots on the wide, very arid, southern plain of barren limestone. All Hawaiians used the leaves for food wrappings, for fiber, and the baked roots for a sugary food.
  There is also a clear record of the shrub *Cordyline* on June 9, 1786, by Capt. Portlock (1789: 84) who listed with the cultivated products which he saw on an excursion, "the sweet root which is called *tee* by the natives".

Sansevieria

Sansevieria trifasciata Prain
  Native name not recorded.
  Not collected, but observed in cultivation in the gardens by the Robinson home at Kiekie. It is an exotic, imported from Africa.

Amaryllidaceae (Amaryllis Family)

Agave

Agave sp.
  Not collected, but observed in cultivation in the village, Puuwai; and also at Kawaihoa Point. Judd recorded (1932) observing the century plant.

Furcraea

Furcraea foetida (L.) Haw.
Malina haole; Malina
Not collected, but observed at Mokouia and Apana Valleys, established on dry slopes. An exotic, imported from tropical America.

Dioscoreaceae (Yam Family)

Dioscorea

Dioscorea alata L.
D. sativa of Hillebrand and of Forbes, as to the Polynesian vegetable, but not as to the bulbiferous established vine, not of Linnaeus.
Uhi
Not seen by the writer who was informed that it was seldom cultivated now, since it suffered during droughts and needed much hand watering to tide it over the dry spells. There is no collection by Stokes in the Bishop Museum.

At the request of Otto Degener, Mr. Sinclair Robinson furnished him in 1935 a large photograph of the Niihau yam. It shows one large tuber, ovate in shape, nearly twice the size of the leaves, and small aerial bulblets from several lower nodes of the climbing stem, and good foliage of hastate, ovate, acute blades. The sharp wings along the stem are visible, verifying the identity of this esculent species of *D. alata*.

Both the exploring ships and the trading vessels that followed them, came to Hawaii in need of supplies. They sought water and traded for food stores, both meat and vegetables. Pigs and chickens were available, and of the vegetables there were taro, sweet potatoes, breadfruit, yam, arrowroot, and *ti*. Of fruits there were bananas and, in the early 1800s, pineapple, watermelon, guava and coconut. The visitors preferred the yam, for its taste and its long keeping qualities, and they sought it in bulk. Yams (*Dioscorea alata*) were produced on all the principal islands, but especially on Niihau. It became a standard practice for ships after visiting or watering in Honolulu or Lahaina, to sail to Niihau to barter for two or three months supply.

The following extracts reveal the extent of the yam trade in Niihau.

Samwell wrote on Feb. 20, 1779, "When we were at Sandwich Islands last Year we got no yams at Atowai (=Kauai) but great plenty at Neehau which is an Island like Morotai and Oronai (=Molokai and Lanai) entirely bare of trees".

He, also, noted on March 10, 1779, "as the Island is thin of Inhabitants, the small patches which are here & there planted with Yams, Taro, and sweet potatoes afford a sufficient supply for them, while large plains of fine land is suffered to lie waste... We procured Yams enough here to serve the 2 ships for bread about six weeks which falls short of the supply we expected." This was on Capt. James Cook's visit on his third voyage.

Capt. James Colnett of the Prince of Wales was at Yam Bay, Oneehow, Jan 1-March 20, 1788, and wrote (p. 18), "During the afternoon several small canoes came off & brought as many yams as they could hold,..."

On June 8, 1786, Capt. Portlock of the *Queen Charlotte* wrote (p. 83) "No sooner were we moored, than several canoes visited us, bringing yams, sweet potatoes, and a few small pigs, for which we gave them in exchange nails and beads." On June 9, he noted (p. 83) "Early the next morning we were surrounded by canoes, who brought a plentiful supply of yams, and some sugar cane. The island appears well cultivated; its principal produce is yams." Then (p. 90), "By this time we had preserved near ten tons of fine yams and Captain Dixon had got about eight tons on board the *Queen Charlotte*."

On a return visit to Yam Bay on January 17, 1787, he wrote (p. 184), "The country seemed poorly cultivated, and Abbenooe told me, that since we took our stock of yams in, the people here in great measure neglected the island, barely planting enough for their own use;..."

Capt. William Robert Broughton of HM sloop *Providence* touched at Yam Bay on February 19, 1796, and reported of (p. 46) bartering for yams.

On July 28-31, 1796, he again was at Yam Bay, trading for yams and this time (pp. 74-80) he sent ashore a party on July 30th in a cutter, and they had trouble. The mate and the botanist, Alexander

Bishop, were attacked by the Hawaiians. Two armed marines were stabbed and killed, and three muskets were lost. In reprisal the crew burnt or destroyed 16 canoes.

Capt. John Turnbull on a world voyage, 1800-1804, landed at Niihau and wrote (p. 221), "the native flocked about us, furnishing abundance of yams at a very moderate rate;..."

Capt. F.W. Beechey of *H.M.S Blossom* stopped at Oneehow and wrote (pp. 234-5), "we were disappointed in the expected supplies; not from their scarcity but in the consequence of the indolence of the natives...We took aboard as many yams as the natives could collect before sunset, and then shaped our course for Kanschatka."

In 1947 and 1949 on the writer's two visits he spent 12 long days on horseback exploring the island, north to south, east to west, with a competent guide. As mentioned he saw no large fertile pockets with vegetable mould. There may be some of these, but they are few and remote; too few, it seems to produce shiploads of yams. The some 300 inhabitants in the late 1700s and early 1800s lived in small settlements mostly scattered along the western, leeward shore. On short notice they would put out to a ship with canoe load after canoe load of yams. Where they grew these yams and how they produced them in such quantity is a mystery.

There are no good accounts of the early agricultural practices on Niihau, and the present writer does not understand how on Niihau there could have been such a large, ready supply of yams. The island is arid. Its leeward shore is low, rather flat, composed largely of elevated coral rock, level basalt flows, sand, or thin dry soil. Hawaiians planted yams preferably in deep, rich soil in gulches or on the edge of native forests (rarely in open garden patches). The growing vines climbed on, or even overtopped the adjacent trees. No such localities or habitats are today evident. Perhaps there were some on the mountain when it was forested, but such places were remote from the anchorage at Yam Bay (which seems to be Nonopapa).

Handy (1972: 180) said that, "Niihau once produced quantities of yams, which were planted presumably in the large pocket in the elevated coral at the southeast of the island, the bottoms of which are said to be filled with vegetable mould". During his visit in 1931 Handy did not see these fertile pockets.

Musaceae (Banana Family)

Musa

Musa paradisiaca subsp. *normalis* Ktze
Not seen recently on the island. It is listed here on the basis of historic records. From Capt. Cook's voyage, Samwell recorded (1967), "Saw a few small plantations of Sugar Canes & Plantains and two or three Palm trees,..."

Dicotyledones

Casuarinaceae (Casuarina Family)

Casuarina

Casuarina littoralis Salisb.
C. equisetifolia L.
Paina (= the word pine, Hawaiianized)
Cultivated tree, 15m. x 30 cm., Kiekie, 50 ft. alt., *St. John 23,658*. This tree had usually large cones, 15-25mm. long, 11-18 mm. in diameter, cylindrical in outline and truncated. It was also observed at Kalanihale and recorded by Judd as ironwood and by Stearns (1945: 6) under its scientific name.

Piperaceae (Pepper Family)

Peperomia

Peperomia leptostachya Hook. & Arn.
P. candollei St. John, Occas. Pap. Bish. Mus. 19(14): 6-8. *pl. ii.* 1931. Described from the Stoke's Niihau collection.
'Ala'alawainui
Foot of mountain on west side, *Stokes*; on shaded basalt ledges, Mokouia Valley, 400 ft. alt., *St. John 23,593*. Also observed at Haao Valley.

Yuncker in his monograph well demonstrated that the Niihau collection is not well separable from the common Hawaiian and wide ranging Pacific species *P. leptostachya*.

Moraceae (Mulberry Family)

Key to the Genera

Flowers inside the hollow stem tip, which becomes a fleshy fruit (fig). *Ficus*
Flowers not so.
    Female flowers on a fleshy receptacle; fruit 25 mm. in diameter.
    Broussonetia
    Female flowers uniting to form a fleshy fruit 8-20 cm. in diameter. *Artocarpus*

Artocarpus

Artocarpus altilis (Parkins. ex Z) Fosb.
    A. incisus L.f.
    'Ulu

One mile northeast of Leahi, 70 ft. alt., in dry limestone sinkhole 15 feet deep, only one old tree 10 m. by 6 dm., projecting 5 m. from the top of the hole, *St. John 22,780*.

    In Former times the *'ulu* grew in several of the sinkholes which were moist at the bottom, but in recent years all the trees have been killed by sheep except this one. A lone sheep, falling into one of these deep sinkholes would be unable to climb out and before starving to death the unfortunate animal would eat everything within its reach, even the bark of the breadfruit trees, thus causing the tree's death.

    This single remaining tree had blades 35-48 cm. long, lobes incised ¾ way to the midrib with the lobes elliptic, acuminate, and overlapping at the middle. The fruit was 18-20 cm. long, ellipsoid, with the rind rather smooth. No varietal vernacular name was known.

    The guide, Kalani, demonstrated that the breadfruit was edible. A fallen fruit so soft ripe as to be squashy, even if moldy on the surface, was found and eaten. Inside, the white pulp was soft and rather slimy, but it had become sweet and was palatable, even tasty and was not unlike a baked, very ripe fruit.

    According to Kawena Pukui (1933: 127) there is on Niihau a bush-like breadfruit differing from the tree. "The low-lying breadfruit is called *kino-o-Haumea*, 'body-of-Haumea', and *na ulu hua i ka hapapa*, 'low-lying like a bush'. It is thought of as a female. The ordinary upright tree is called male and named *ulu ku*, or 'upright breadfruit'."

This breadfruit also is mentioned in the "Mele a Kapi'olani", recorded by Pua Ha'aheo: "Eho'o ke aloha i Niihau ea" etc., and in translation:

"Love goes back to Niihau
To the hidden pool of the pao'o fish,
To the breadfruit that bears close to the ground,
(and) the sugar cane of Halalii that is dug up
Beyond and back lies Nehoa,
A rocky islet in the sea.
The hot sun shines on one's back.
As one turns back to Kauai."

Artocarpus growing in sinkhole, Leahi.

Handy (1972: 152) mentions this tradition and continues: "An informant on Kauai told of an *'ulu* growing on rocky places on Niihau, which she termed *'lu ne'e i ka papa* or *'ulu kupua*. She described it as having leaves and fruit like the breadfruit, but as being inedible. Mrs. Pukui says Niihau *meles* mention *'ulu ne'e i ka hapapa*, the breadfruit that creeps on the rocks. This is the same as the *'ulu hua* mentioned above, and it grows only on Niihau."

An anonymous visitor to Niihau (*The Friend*, Oct. 1870, p. 94) wrote "...I was shown some subterranean caves, in which were growing with great luxuriance the breadfruit, *hau*, and many other trees. This was within a stones throw to the ocean." It is probable the visitor should have called the spots sinkholes, for dark caves cannot produce healthy trees. The sinkholes such as the one shown to the present writer, are shady and moist below, but the top is open to the sun, forming a habitat excellent for the growth of trees or other plants.

Niihau does not have a distinct kind of breadfruit, a vine-like plant that creeps on the rocks. This is a concept of people who got their idea from literature, and who have never seen the Niihau sinkholes in the elevated coral rock, or the trees growing therein. The Niihau *Artocarpus* trees produce both male and female flowers, like those of other regions, and they are not unisexual, as stated by Pukui. The fruit is edible.

Broussonetia

Broussonetia papyrifera (L.) Venten.
Wauke
On basalt cliff of small gorge, Haao Valley, 300 ft. alt., 1 m. shrub, *St. John 23,638*. Thicket in gulch, north side of Halulu Lake, 50 ft. alt., 5 m. shrub, *St. John 22,739*.

This is the shrub formerly used by the Hawaiians for the manufacture of bark cloth or *kapa*. It is native to eastern Asia, but was carried along and cultivated by the Pacific Islanders. In 1947 Mrs. Mary K. Pukui told the writer of the old accounts indicating that the natives did not have *wauke* on Niihau and hence dressed in *makaloa* mats. From the fact that *wauke* now grows on the island, it may well have done so in aboriginal times. Due to the aridity, however, it would have made slow growth and may not have produced enough bark to meet the clothing demands of the people.

Ficus

Ficus carica L.
Piku (= fig, Hawaiianized)
Not collected or observed in cultivation at either Puuwai or Kiekie.

Proteaceae (Silky Oak Family)

Grevillea

Grevillea banksii R. Br.
Oka pua ʻulaʻula (=oak, flower, red); *Lauʻau kepani* (=medicine, Japanese)
South ridge of Mokouia Valley, 1000 ft. alt., dry upper slopes, planted, trees 6 m. by 1 dm., *St. John 22,814*. Also reported by Stearns (1947: 6), "The chief success in reforestation of the uplands has been accomplished with silk oak (*Grevillea robusta*), haole koa (*Leucaena glauca*), and Grevillea (*Grevillea banksii*)." This is the tree unique in having poisonous corolla-like calyces that may on contact cause dermatitis. Introduced from Australia.

Grevillea robusta A. Cunn. ex R. Br.
Oka (= oak, Hawaiianized)
Ridge, 1 mile east of Kaeo, 750 ft. alt., planted, trees 8 m. by 2 dm., *St. John 22,797*. See passage above under *G. banksii*. Introduced from Australia.

Santalaceae (Sandalwood Family)

Santalum

Santalum ellipticum var. *littorale* (Hbd.) Skottsb.
Northern slope, Kawaihoa Cove, Jan. 15, 1977, *C. Christensen 109*. Leaves orbicular, fleshy, like the plant formerly called *S. cuneatum* var. *laysanicum* Rock.

Chenopodiaceae (Goosefoot Family)

Chenopodium

Leaves sticky-hairy.	*C. carinatum*
Leaves dry and mealy.	
Shrubs; herbage often ill-smelling; seeds dull edged.	*C. oahuense*
Herbs; herbage inoffensive ; seed sharp edged.	*C. album*

Chenopodium album L.
    Pakapakai
    A weedy plant, introduced from Europe. Abundant in corral, also on grassy flats, Kii, 15 ft. alt., *St. John 23,606*; on sand flats near shore, Nonopapa, 20 ft. alt., *St. John 22,771*. Also observed at Apana Valley and Kamalino. A weedy plant introduced from Europe.

Chenopodium carinatum R. Br.
    No native name.
    Weed in grassy opening between *Prosopis* trees, Paniau, 800 ft. alt., *St. John 23,588*. Also observed at Apana Valley, Halulu Lake and Halalii Lake. This is a rank weed, introduced from Australia.

Chenopodium oahuense (Meyen) Aellen
    C. sandwicheum Moq.
    C. oahuense f. *macrospermum* Aellen
    C. oahuense f. *microspermum* Aellen
    Alaweo huna (huna = secret, *alaweo* may be a variant of *aweoweo*, another name for this species)
    Kaali, *Stokes*; on mud flat ½ miles inland, common, Palikoae, 10 ft. alt., *St. John 22,713*.

The two forms described by Aellen are here reduced to synonymy under the species. The forma *macrospermum* was described as having the seed small, ¾ mm. in diameter; pericarp thin without color. In all he cited five collections. Before the present writer there were many times that numbers of specimens and no differences in the pericarp were recognized. There are differences in size of seed, but there is no clear division into two groups, one with large and one with small seeds. Instead, there is a continuous gradation between the large and the small. Finally, from a single plant, one can find the large and the small sized seeds. Hence, it is deemed correct to reduce both of the forms to the synonymy of the species.

The species is widely distributed in the arid sections of the Hawaiian Islands, and the writer has met it frequently, and knows it well by sight and by smell. Its herbage has a strong, pungent, fishy, or almost fetid odor. The collection *St. John 22,713* seems in every morphological detail exactly this species, yet, in the field it was noted the herbage was only slightly aromatic, not unpleasant. One other collection, *Fosberg 10,168* from the 1859 flow, North of Kona, Hawaii, has the data, "no noticeable odor". On every morphological character these seem exactly the species, yet from the lack of the strong characteristic oil, they may represent physiological variants.

Atriplex

Atriplex semibaccata R. Br.
Sand near beach, Kii Landing Feb. 12, 1977, *C. Christensen 140*. Mud flat, Puuwai, 10 ft. alt., Feb 12, 1977, *C. Christensen 148* and *153*.

Amaranthaceae (Amaranth Family)

Amaranthus

Amaranthus viridis L.
Pakapakai
A weed, introduced from the tropics, observed, but not collected.

Nototrichium sandwicense var. *niihauense* St. John, Pacif. Sci. 13: 166-168. *fig. 4* . 1959.
First valley west of Kaali Cliff, top of steep basalt talus, 100 ft. alt., August 16, 1947, *St. John 22,830*, the type collection. Across from the Puu'uala, Kaalai Cliffs, 200 ft. alt., Feb. 12, 1977, *C. Christensen 174* and *175*.

Nyctaginaceae (Four o'clock Family)

Boerhavia

Boerhavia diffusa var. *sandwicensis* Heimerl
Anena
One mile northeast of Kiekie, 20 ft. alt., trailing on sand near shore, flowers white, *St. John 22,791*; Kii, *Stokes*; the latter determined and listed as *B. diffusa* by Forbes.

Boerhavia diffusa var. *pseudotetrandra* Heimerl
Anena
North Kona Cliff, talus, *Stokes*; this was listed by Forbes as *B. tetrandra*.

Bougainvillea

Bougainvillea spectabilis Willd.
Pua pepa (= flower of paper)
Observed in cultivation by the houses in the village of Puuwai, but not collected.

Batidaceae (Batis Family)

Batis

Batis maritima L.
'Ae'ae mahu (= Lycium, weak)
Exsiccated bed of Halalii Lake, 50 ft. alt., abundant on salty flat, shrubs 1m. tall, *St. John 22,734*.

Aizoaceae (Carpetweed Family)

Sesuvium

Sesuvium portulacastrum (L.) L.
'Akulikuli
On spray covered coral ledges, Kaununui, 5 ft. alt., flowers lavender, *St. John 22,716*; north base of Kawaihoa Point, wet seep on cliff, *St. John & Tuthill 22,784; 22,787*. Also observed at Mokouia Valley, Kiekie, Nonopapa, Leahi, Halulu Lake and Poooneone. It is common along the sandy and rocky seashores and by the muddy margins of the inland saline lake beds. This species was listed by Forbes (1913: 21), but the specimen so labeled in his hand is *Bacopa monnieria*.

Portulacaceae (Purslane Family)

Portulaca

Petals dark rose-colored; leaves slender, 1/16 in. wide.
                                                              P. cyanosperma
Petals yellow; leaves inverted egg-shaped.   *P. oleracea*

Portulaca cyanosperma Egler
'Akulikuli
Back of dune, Kaununui, 15 ft. alt., flowers magenta, *St. John 22,719*.

Now common throughout the lowlands, especially in sandy places. To the observer it appeared indigenous, although it was only recently described as a native of Kauai and Lehua, but he accepts the judgment of the natives of Niihau. Some stated that it was introduced by horses brought from Kauai to Niihau about the year 1942. Others said that it appeared about the year 1932 at the water hole at Kalanihaale Landing, and thought it had been brought there from Lehua Island by birds. The distance from this offshore mountainous islet is only one mile. As it is the home or nesting site of great quantities of sea birds, and as Lehua Islet, with its steep, shelving, rocky slopes lacks standing water, it is natural for the birds to visit the nearby water hole on Niihau. They may well have transported the plant, or its seeds. It is also abundant on the nearby Barking Sands on Kauai, which may well be its original home. In any case, it must be considered a recent adventive on Niihau.

Portulaca oleracea L.
   '*Akulikuli*

One half mile northeast of Kiekie, 505 ft. alt., weed by path, *St. John 22,793*. Also observed at Kaununui and Puuwai. Its occurrence was as a weed in the village and as an occasional ruderal weed by paths and roads.

Caryophyllaceae   (Pink Family)

Schiedea

Schiedea amplexicaulis Mann

This species has been collected only once. The type is *Rémy 548 bis*, in the Gray Herbarium. Its label gives as locality "Kauai ou Nihau" and the date 1851-55. The duplicate specimen in the Paris herbarium bears identical data. This species has been accepted and its description amplified by the monographer, Sherff (1945: 322). Since many of the species of *Schiedea* are denizens of the lower scrub or dry lower forests, this species may well have been endemic to Niihau. It has never been collected again, and it is now patently extinct.

Menispermaceae (Moonseed Family)

Cocculus

Cocculus ferrandianus Gaud.
 Foot of plateau, southeast, *Stokes*; two sheets with firm ovate leaves, one stem with denuded pedicels. Observed in sterile condition by the writer at Kaununui. It is now very rare.

Lauraceae (Laurel Family)

Cassytha

Cassytha filiformis L.
 Kauno'a
 Kiekie, *Stokes*; south half of island, *Stokes*; edge of thicket, in dunes, Poooneone, 15 ft. alt., dense tangle on various hosts, *St. John 22,740*. Also observed at Kiekie. It is common near the seashore.

Persea americana Mill.
 Cultivated as a fruit tree, in village of Puuwai and at the Robinson home at Kiekie.

Papaveraceae (Poppy Family)

Argemone

Argemone glauca Pope
 Puakala
 Foot of plateau, southeast, *Stokes*; on dry hill, Kawaewae, 270 ft. alt., plant 2 m. tall, petals white, *St. John 22,727*. Not seen elsewhere.

Capparaceae (Caper Family)

Key to the Genera

Fruit fleshy, not splitting; leaves entire.   *Capparis*
Fruit dry, splitting open; leaves 5-lobed.   *Gynandropsis*

Capparis

Capparis sandwichiana DC.
Pilo; also *puapili*, the latter being the name formerly used by the common people, while the chiefs called it *maiʻa a Maui*.
Foot of plateau, Southeast, *Stokes*; southern end of Halalii Lake, 50 ft. alt., on coral rock ledge, branches 1 m. long, arching, flowers white, fading pink, none fully open at noon, *St. John 22,733*; on tuff rocks, Kawaihoa Point, depressed shrub, 200 ft. alt., *St. John 23,614*. Also observed at Keawanui Bay, Kamalino and Leahi. Occasional on high coral ledges in the lowlands, out of reach of grazing animals.

Gynandropsis

Gynandropsis gynandra (L.) Briq.
Niihau, collected, apparently by W. T. Brigham, distributed as by *H. Mann & W. T. Brigham*, without number. The specimen of this in the Bishop Museum was determined by collectors as *Cleome sandwicensis* Gray, and it was recorded under this wrong name by Mann (1867: 149). It is a pantropic weed, introduced to the Hawaiian Islands. It has not been collected on Niihau by any recent collector.

Cruciferae (Mustard Family)

Key to the Genera

Pods of 2 rounded lobes which fall separately. *Coronopus*
Pods long and narrow. *Lepidium*

Coronopus

Coronopus didymus (L.) Sm.
 ʻIhiʻihi-i-one
 Kaali, *Stokes*.

Lepidium

Plant without hairs; blades lance-shaped; stamens 2.
L. o-waihiense
Inflorescence hairy; blades obovate; stamens 6. *L. virginicum*

Lepidium o-waihiense C. & S.
Collected in 1826 by Lay and Collie on the island, which they called Oneeheow or Nihow. Not recently collected there, so probably now extinct on Niihau.

Lepidium virginicum L.
Weed in grass by corral, Kanalo Valley, 20 ft. alt., *St. John 23,596*. A weed introduced from North America.

Rosaceae (Rose Family)

Rosa

Rosa sp.
A species of rose, observed in cultivation in 1947, not collected or determined as to the species.

Leguminosae (Fabaceae) (Pea Family)

Key to the Genera

Petals equal, in bud the edges touching.
  Stamens more than 10.
    Pods spiral or strongly twisted. *Pithecellobium*
    Pods straight. *Albizzia*
  Stamens 5-10.
    Anthers not gland-tipped.
      Pods straight; tree; leaves not moving. *Leuceana*
      Pods curved; low herb; leaves and leaflets sensitive, moving when touched. *Mimosa*
    Anthers gland-tipped; tree with thorny branches. *Prosopis*
Petals unequal, in bud overlapping.
  Petals very unequal, the upper petal the outermost.

Pods when ripe separating into 1-seeded joints. *Desmodium*
Pods splitting lengthwise into 2 valves, or not splitting.
    Leaves or leaflets without stipels.
        Leaves with 5 or more leaflets, ending in a tendril;
            pods black, not inflated. *Abrus*
        Leaflets 3 (or 1), the margins unlobed; pods inflated.
            Crotolaria
    Leaves or leaflets with stipels.
        Pods not splitting open.
            Leaves even pinnate. *Sesbania*
            Leaves odd pinnate.
                Pods rounded in cross section, curved.
                    Indigofera
                Pods flat, straight. *Tephrosia*
        Pods splitting into 2 valves.
            Stalk of leaflets with glands; seeds red. *Erythrina*
            Leaflets with stipels instead of glands; seeds not red.
                Style bearded below stigma; pods linear.
                    Phaseolus
                Style not so; pods oblong. *Canavalia*
Petals slightly unequal, the upper petal innermost in bud.
    Leaves bipinnate. *Caesalpinia*
    Leaves pinnate.
        Flowers with a strongly dilated disk; trees. *Ceratonia*
        Flowers not so; shrub. *Cassia*

Abrus

Abrus precatorius L. f. *precatorius*
    Pukiawe ʻulaʻula (= *styphelia*, red)
    Southern half of island, Jan. 1912, *Stokes*; dry thicket, 20 ft. alt., *St. John* 22,766. Seeds red with a black spot.

Abrus precatorius f. *lutiseminalis* St. John, Pacif, Sci. 13: 168. 1959.
   Pukiawe lenalena (=*styphelia*, yellow)
   In scrub on dry limestone flat, Nonopapa, 20 ft. alt., Aug. 13 , 1947, *St. John 22,768*. This is the type collection of the form which has its seeds entirely pale yellow. It is evidently a local variant, evolved from the introduced species.

Albizzia

Albizzia lebbeck (L.) Benth.
   'Ohai
   Dry hillside, Halulu Lake, 50 ft. alt., trees 7 m. by 2 dm., flowers greenish, Aug. 12, 1947, *St. John 22,725*. It was also observed by the writer, as a planted tree, at Kaununui and Kamalino. It was observed in 1932 and recorded as the "siris tree" by C. S. Judd (1932: 5-9).

Caesalpinia

Caesalpinia bonduc (L.) Roxb.
   C. jayabo sensu Rock, Legum. of Hawaii 103-105. 1920.
   Kakalaioa; *kinikini* (=marbles, a name given to the plants as its hard, gray seeds were used as marbles by the children)
   Open woods on shore, Halulu Lake, 50 ft. alt., climbing 4 m. over trees, *St. John 22,724*; 1 mile northeast of Leahi, dry thicket on coral rock, 70 ft. alt., *St. John 22,777*. Also observed at Apana Valley, Kaeo and Halalii Lake.

Canavalia

Canavalia pubescens Hook & Arn.
   Paunu (= thick growing)
   Oneeheow (= Niihau Island), *Lay & Collie,* Beechey Voyage, (Herb. Kew), the type specimen of this species; foot of mountain on west side, Jan. 1912, *J.F.G. Stokes*; small gorge, Haao Valley, 300 ft. alt., long vine, forming thicket, flowers madder, pod black, *St. John 23,637*. Endemic to Niihau.

Cassia

Cassia leschenaultiana DC.
Chamaechrista leschenaultiana (DC.) Degener
Ti Kepani (= the tea of the Japanese)
On dry limestone flat, Nonopapa, 20 ft. alt., *St. John 22,772*. Also observed at Mokouia Valley and Paniau. A weed, introduced from India.

Cassia occidentalis L.
Miki palaoa
Foot of plateau, southeast corner, 1912, *J.F.G. Stokes*. Also observed by the writer at Halulu Lake and Kamalino. It is an introduced, pantropic weed.

Ceratonia

Ceratonia siliqua L.
Uhini
One mile east of Kiekie, planted in a gulch, 50 ft. alt., *St. John 22,750*. This is the cultivated "carob" or "St. John's Bread", introduced from the Orient.

Crotalaria

Leaflets 3.
Pods without hairs; leaflets 3-3.5 cm. wide, above without hairs.
    C. pallida
Pods and leaflets hairy; leaflets 1.5-3 cm.   *C. incana*
Leaves simple, 7.5-15 cm. long, silky above.   *C. spectabilis*

Crotalaria incana L.
Kolomona (= Solomon, the same name applied on the larger islands to *Cassia glauca* and *C. laevigata*), the bright yellow flowers of which are likened to "Solomon in all his glory".

Common in grasslands and pastures, Paniau, 1,100 ft. alt., *St. John 23,584*. Also observed at Kii and Kaali Spring and Cliff. It was introduced from tropical America.

Crotolaria pallida Aiton
 C. mucronata Desv.
 Kolomona (= Solomon)
 Not collected, but common, and observed at Kii, Kaali Spring and Cliff, Mokouia Valley, Puuwai, Apana Valley, Halalii Lake and Kamalino. Introduced from the tropics of the Old World.

Crotolaria spectabilis Roth
 Kolomona (=Solomon)
 In dried mud, bed of old Makanikahau Reservoir, Apana, 400 ft. alt., *St. John 23,632*. Introduced from India.

Desmodium

Pods notched on lower edge; leaflets broadest near the tip.
 D. trifolium
Pods notched on both edges; leaflets broadest near the base.
 D. uncinatum

Desmodium triflorum (L.) DC.
 This was recorded by C.N. Forbes (1913: 22) as *Meibomia triflora*. The specimen, collected by Stokes, has not been found in the Bishop Museum, but there seems no reason to doubt this record. It was introduced from Asia or Africa.

Desmodium uncinatum (Jacq.) DC.
 Meibomia uncinata (Jacq.) Ktze.
 Pilipili'ula; pilipili; 'olena
 South half of island, Jan. 1912, *J.F.G. Stokes*. Observed by writer at Kaali Cliff, Mokouia Valley, Paniau, Apana Valley and Halalii Lake. Introduced from tropical America.

Erythrina

Seeds red.
  Flowers orange, yellow or scarlet.
    E. sandwicensis var. sandwicensis f. sandwicensis
  Flowers yellowish-green.   *E. sandwicensis* f. *lutea*
  Seeds dull yellow; corollas pale green.
    E. sandwicensis var. *luteosperma*

Erythrina sandwicensis Degener var. *sandwicensis* f. *sandwicensis*
  Wiliwili
  Foot of plateau, southeast, Jan. 1912, *Stokes*. Reported by Capt. Portlock, of the "Queen Charlotte" (1789: 85); and by Stearns (1947: 6). The writer noted it as common, observing it at Keawanui Bay, Halulu Lake, Kamalino, Mokouia Valley and Kapaka Valley.

Erythrina sandwicensis f. *lutea* St., John, Pacif. Sci. 13: 171. 1959.
  Wiliwili
  Rocky dry gulch Apana Valley, 400 ft. alt., *St. John 22,806*. A form with yellowish-green corollas and red seeds. This is the holotype and the type locality.

Erythrina sandwicensis var. *luteosperma* St. John, Pacif. Sci. 13: 171. 1959.
  Wiliwili lenalena
  Thicket on dry flat, Nonopapa, 20 ft. alt., *St. John 22,769*. This is the type and the only known collection.

Indigofera

Indigofera suffruticosa Mill.
  Inikoa (= a Hawaiianization of indigo)
  North Kona Cliff, Jan. 1912, *J.F.G. Stokes* (as *I. anil* L.); Kawaewae, 75 ft. alt., common on dry hills, *St. John 22,729*. It is abundantly established and was also observed at Kii, Kaali Cliff, Mokouia Valley, Halulu Lake, Halalii Lake, Kamalino and Kawaihoa Point. Introduced from South America.

Erythrina sandwicensis tree in Apana Valley, 1947
(*photo by H. St. John*)

Leucaena

Leucaena leucocephala (Lam.) de Wit
L. glauca (L.) Benth.
Koa; koa mahu (= imitation koa)
First observed in 1945 by Stearns (1947: 6), as "successful in reforestation". Thicket on dry flat, Nonopapa, 20 ft. alt., 5 m. shrub, *St., John 22,765*. Also observed, but not collected, at Kaali Cliff, Kapaka Valley, Paniau and Kaeo. It is occasional on the flats of the western shore as far south as Nonopapa, but on the uplands it is common, and in the region of Paniau and the upper Mokouia Valley it has grown to form a forest 6 meters tall and so dense in stand that

in many places one cannot walk between the trunks. It flowers and fruits, but throughout the long dry season it is largely deciduous and bare. It is useful as stock feed. Introduced from tropical America.

Mimosa

Mimosa pudica var. *unijuga* (Duchass. & Walp.) Griseb.
   Hilahila, pua hilahila (= flower, bashful)
   Ridge 1 mile east of Kaeo, 750 ft. alt., by trail in pasture, *St. John 22,798*. It is common and introduced from tropical South America.

Phaseolus

Phaseolus lathyroides L.
   Uhiuhi (On the other islands this vernacular name is applied to the endemic leguminous tree, *Mezoneuron kavaiense*).
   Haao Valley, 200 ft. alt., weed in pasture, *St. John 22,805*. Introduced from tropical America.

Pithecellobium

Pithecellobium dulce (Roxb.) Benth.
   'Opiuma
   Mokouia Valley, 600 ft. alt., 8 m. tall, tree planted by trail in *Prosopis* thicket, *St. John 23,595*; Kawaewae, 75 ft. alt., dry hillside, *St. John 22,730*. Also observed, but not collected at Halulu Lake. It was introduced from tropical America.

Prosopis

Prosopis pallida (Humb. & Bonpl. ex Willd.) Kunth.
   P. juliflora, sensu Hawaiian authors, not of (Sw.) DC.
   Kiawe
   Listed (as *P. juliflora*) by Forbes (1913: 22), but neither Stokes nor any other botanical visitor has preserved a specimen. Stearns observed (1949: 6) "*Kiawe* or algaroba *(Prosopis chilensis)* spread over rocky ledges and the lowland until it became a pest and now has to be cleared. Transpiration by *kiawe* trees dried up several small but

Savannah with *Prosopis*, northeast from Kapaka, 1949
(*photo by H. St. John*)

valuable springs and caused several water holes to become brackish". The writer quotes this, but in relation to springs drying up, or water holes turning brackish, does not confirm it.

The writer observed, but did not collect the tree at Kii, Kaali Cliff, first valley west of Kaali Cliff, Kapaka Valley, Mokouia Valley, Paniau, Keawanui Bay, Kauninui, Puuwai, Kiekie, Nonopapa, Halulu Lake, Halalii Lake, Kalaalaau Valley, Poooneone, Kawaewae, Kamalino, Leahi and Kaumuhonu Bay. Actually the tree is the dominant plant of the island. It occurs sparingly on the flats north of Kaali. On the flats from Halulu Lake to the south end of the island it is widely distributed, but it makes a sparse stand. On the remaining lowlands forming a coastal plain from Mokouia to Kiekie it is abundant, making a dense tree cover. On the mountainous upland it ascends the leeward valleys and their slopes to 900-1,000 feet altitude, where it is replaced by grasslands. It furnishes feed, both herbage and nutritious pods for the grazing stock. The tree was introduced from South America.

Sesbania

Sesbania tomentosa Hook. & Arn.
'*Ohai a Papiahuli* (= the '*ohai* of *Papiahuli*)
Headland W. of Kaumuhonu Bay, near crest of grassy headland, 160 ft. alt., *St. John 22,781*.

This species is now approaching extinction on Niihau, there being only a few erect shrubs 1 meter tall, on the grassy headlands at the extreme south end of the island. Forbes (1913: 22) listed *Sophora tomentosa* as native to the island. The latter is a species of the tropical shores of both hemispheres and is widely distributed in the Pacific, but it is not native to the Hawaiian Islands. There it is of recent introduction, and is rare and only in cultivation. It is unknown now on Niihau. No specimen of this *Stokes* collection is now to be found in the Bishop Museum, nor is there a record of it ever being listed in the card catalog of the herbarium. It is suggested that by oversight Forbes wrote *Sophora tomentosa* Hook & Arn. when he should have written *Sesbania tomentosa* Hook. & Arn.

Tephrosia

Tephrosia purpurea (L.) Pers.
T. piscatoria (L.) Pers.
Hola

South half of island, Jan. 1912, *J.G.F. Stokes*; west slope in grassland, Kawaihoa Point, 150-300 ft. alt., *St. John 23,612*. This is the famous fish-poison plant.of the Hawaiians and other Polynesians. Not finding it at first, the writer asked for it. The natives knew of it, said it formerly grew on the island, was no longer used for fishing, and in fact was extinct. It was one of the very few native plants of aboriginal use that the guide could not go to.

Oxalidaceae (Wood Sorrel Family)

Oxalis

Oxalis corniculata L. var. *corniculata*.
Kaali, Jan. 1912, *J.G.F. Stokes*. There is also an early record of *O. repens* Thunb. for Niihau, as collected by Lay and Collie, published by Hooker and Arnott (1832: 80). They state, under the heading *Oxalis repens*, that "Our specimens, found at Oneeheow, are neither in flower nor in fruit, and we should have referred then to *O. corniculata*, but Gaudichaud having mentioned *O. repens* as a native of the Sandwich islands, and not *O. corniculata*, we have retained the above name". In response for a request for verification of the above, Sir Edward Salisbury, Director of the Royal Botanic Garden, Kew, replied in a letter of Sept. 12, 1952, "The only Beechey specimen that we have been able to trace in the Kew herbarium came from Gambier's Island (Society Islands) and this is *O. corniculata* L."

Zygophyllaceae (Tribulus Family)

Tribulus

Tribulus cistoides L.
Nohu
Kii, Jan. 1912, *J.F.G. Stokes*; coral sand by beach, Nonopapa, 20 ft. alt., *St. John 22,761*. Also observed by the writer at Kalanihale, Keawanui Bay, Poooneone, Kamalino, Leahi and Kawaihoa Point. It is generally distributed and abundant along the sandy shores.

Rutaceae (Rue Family)

Citrus

Citrus sinensis (L.) Osbeck
'Alani, or *'alani Hawaii*
A cultivated tree, observed at Mokouia Valley, Puuwai, Haao Valley and Halulu Lake. Introduced from tropical Asia.

Citrus grandis (L.) Osbeck
    'Alani pake (= the Chinese orange)
    Observed in cultivation at Leahi. Introduced from southeast Asia.

Murraya

Murraya paniculata (L.) Jack
    Observed in cultivation in the village, Puuwai. Shrub, introduced from Asia.

Meliaceae   (Mahogany Family)

Melia

Melia azedarach L.
    'Inia (= India)
    South half of island, Jan. 1912, *Stokes*. Observed by the writer at Keawanui Bay, Puuwai, Apana Valley, Kiekie and Kawaewae. It is a planted tree, introduced from India.

Euphorbiaceae   (Spurge Family)

Key to the Genera

Trees.
    Fruit husk smooth, containing a nut; calyx and corolla present.
                                                Aleurites
    Capsule spiny, with bean-like seeds; only the calyx present.
                                                   Ricinus
Herbs or shrubs.
    Flowers in fused bracts; perianth lacking; sap milky.
                                                  Euphorbia
    Flowers with calyx and corolla; leaves lobed; fruit a capsule.
                                                  Jatropha

Aleurites

Aleurites moluccana (L.) Willd.
Kukui
 Halulu Lake, one tree in gulch, said to be recently introduced from Kauai, *St. John 22,720*. Also observed at Haao Valley, Apana Valley and Kawaewae. Capt. Nathaniel Portlock in June 1786 put his ship the *King George* into Yam Bay (= Nonopapa) seeking yams and water. Inland he found a few scattered trees, one variety, (1789: 85) "with nuts growing on them like our horse chestnut. These nuts, I understand, the inhabitants used as a substitute for candles, and they give a most excellent light." There is no doubt that this account referred to the candlenut, or *Aleurites*. Nuts without the husk are also common as drift on the windward shore, doubtless floating from Kauai. They never germinate on the beaches, and the ones examined were either empty or decayed or putrid. The species is native to southeast Asia, but was introduced to Hawaii several centuries ago by the immigrating Polynesians.

Euphorbia

Herbs.
 Maritime smooth plants with heart-shaped leaves.   *E. degeneri*
 Terrestrial plants with elliptic or ovate leaves.
  Leaves all nearly the same size, with brownish spot near center; plants hairy.
  Leaves variable, plants smooth.   *E. hirta*
   Upper leaves crowded and much larger than the lower leaves.   *E. geniculata*
   Upper leaves or bracts smaller, opposite; lower leaves larger, whorled or alternate.   *E. peplus*
Shrubs; leaves elliptic.   *E. celastroides*

Euphorbia celastroides Boiss. var. *celastroides*
 Chamaesyce celastroides (Boiss). Croizat & Degener
 'Akoko
 Niihau, in 1851-55, *J. Rémy* (Paris), and fragment (BISH). This is the type collection. In the Paris collection this is listed as *Rémy 595*.

Kaali, Jan. 1912, *J.F.G. Stokes*, two sheets; dry basalt talus slope, Kaali Cliff, in *Leucaena leucocephala* scrub, 750 ft. alt., *St. John 22,836*; on north base, windward side, at wet seep on sea cliff, Kawaihoa Point, 20 ft. alt., *St. John & L.D. Tuthill 22,785*. Sherff in his monograph (1938: 26) describes the species, or as it now should be called *E. celastroides* var. *celastroides*, as with the "Cymes axillary 1-5-cephalous." Our abundant collection, no. *22,836*, has them 5-15-cephalous. As now treated, this is a large species with numerous varieties, but the original and type collection is from Niihau.

Euphorbia celastroides Boiss. var. *Stokesii* (C. N. Forbes) Sherff
   E. stokesii C.N. Forbes
   'Akoko

Kii, near the beach, Jan. 1912, *J.F.G. Stokes*, the type collection, two sheets. In tuff ridge slope, Kawaihoa Point, 200 ft. alt., *St. John 23,615*. Forbes described this as a new species because of its fleshy, obovate leaves. Sherff placed it (1938: 26-27) as a variety of *E. celastroides*, characterized by "its more slender branchlets, the often present low terminal foliar lobe (manifest because of the two additional emarginations), and by its greenish capsules". Besides the broad leaf shape, Sherff stressed the terminal foliar lobe. Forbes (1913: 26) had referred to the same structure as, "Leaves... obovate, often emarginate or turbinate..." Forbes' photograph of the holotype shows 57 leaves, of which 6 have detectable lateral notches. This same holotypic sheet now has attached, or in a pocket, 49 leaves of which 11 are lobed. The second or isotypic sheet has either attached or loose in a pocket 38 leaves, of which 10 are lobed. Frequently of a pair of leaves at a node, only one will be lobed. The lobing feature is not constant and does not occur on the majority of the blades. It might have been caused by an injury, whether mechanical or by salt water action, during a young growing stage. In any case, it does not seem to be a fixed characteristic of the plant. Nevertheless, the plant seems well classified as a variety of *E. celastroides*.

Sherff (1938: 26) says "pedicel...usually becoming about 1 cm. long". It is noted that on the *Stokes* collection from Kaali and on *St. John 22,836*, many are only 2.5 mm. long. On the several collections from Nihoa Island there are also many short-stalked ones. It is suggested that a better statement would be: pedicels 2-12 mm. long.

The writer's collection, *St. John 22, 785*, from an exposed, salty sea cliff has leaves intermediate in shape, some of them quite as broad as those of var. *stokesii*. The no. *23,615* has the leaf breadth of var. *stokesii*, but all the leaves are entire.

Euphorbia degeneri Sherff
'Akoko
Kalanihale, ⅛ mile south of Puukole Point, in sand dunes near beach, 15 ft. alt., *St. John 22,712*. Also observed at Kii and Kawaihoa Point. Not common, but native on the sandy shores.

Euphorbia geniculata Ortega
South half of island, Jan. 1912, *J.F.G. Stokes*. Also observed by the writer at Puuwai. It is uncommon, a weed, introduced from tropical America.

Euphorbia hirta L.
  E. pilulifera L.
Lihilihi kahahiaka (= fringe of the morning)
South half of the Island, *J.F.G. Stokes*. Observed by the writer at Kii, Kaali Cliff, Kiekie and Kamalino. It is a weed, introduced from tropical America.

Euphorbia peplus L.
This was listed by Forbes (1913: 23) in his report on the Stokes collection. No specimen of it is now found in the Bishop Museum and it is not listed in the card catalogue as having been inserted in the herbarium. It is a weed of sporadic occurrence in the Hawaiian Islands, and the report of its occurrence may well have been correct. It is an introduction from Eurasia.

Jatropha

Jatropha curcas L.
Kukui pake (= Chinese candlenut)
North of Halalii Lake, planted on knoll, 75 ft.alt., *St. John 22,751*.
In Hawaiian herbal medicine the nuts of the *kukui* (*Aleurites moluccana*) were used as a laxative or purge. The vernacular name in use on Niihau suggests that the violently poisonous seeds of the

Jatropha were used in the same way. As they may be fatally poisonous, it is hoped that this is not a general practice. The species was introduced from tropical America.

Ricinus

Ricinus communis L.
 Koli
 One mile west of Kawaewae, 100 ft. alt., in basalt rocks in open area, *St. John 23,624*. Also observed at Mokouia Valley, Puuwai, Apana Valley, Halulu Lake, Kamalino and Leahi.

Anacardiaceae (Mango Family)

Mangifera

Mangifera indica L.
 Manako (= Hawaiianization of mango)
 Not collected, but observed in cultivation at Puuwai and Kiekie. It has been introduced from India.

Schinus

Schinus terebinthifolius Raddi
 Not collected, but observed in cultivation at Puuwai. It was introduced from Brazil.

Sapindaceae (Soapberry Family)

Key to the Genera

Petals 4; leaves divided; fruit inflated.	*Cardiospermum*
Petals none; leaves simple; fruit winged.	*Dodonaea*

Cardiospermum

Cardiospermum halicacabum L.
C. *microcarpum* Kunth
C. *halicacabum* var. *microcarpum* Bl.
Hale o ka iʻa (= home of the fishes)
Foot of mountain on west side, Jan. 1912, *J.F.G. Stokes*. Observed by the writer at Puuwai.

Plants with fruits 3-4 cm. long have been taken to represent the species. Those with fruits 2 cm. or less in length have been separated as the species, or the variety *microcarpum*. No other significant characters are recorded. The fruits are large, inflated, and bladder-like. The degree of inflation is variable. Among the specimens studied in the Bishop Museum there is no such clear-cut separation into plants with fruits less than 2 cm. long and those with fruits 3-4 cm. long. On the contrary, there is a complete gradation from smaller to larger fruits, without any perceptible break. The var. *microcarpum* is confidently reduced to synonymy. It is introduced from the Old World.

Dodonaea

Dodonaea eriocarpa Sm. var. *obtusior* Sherff f. *obtusior*
D. *viscosa* sensu C.N.Forbes, not of L.
ʻAʻaliʻi

Foot of mountain, west side, Jan. 1912, *J.F.G. Stokes*; Paniau, by *Prosopis* thicket, 1,000 ft. alt., *St. John 23,585, 23,586*; ¼ mile south of Poooneone, 20 ft. alt., inner sand dunes, *St. John 23,648, 23,649, 23,650*; Kawaewae, dry hillside, rare, 100 ft. alt., *St. John 22,728*. Also observed at Apana Valley. There are still remnant patches of this on the upland of the northern half of the island, and it appears to have been an important part of the forest cover formerly existing on that mountainous section.

Rhamnaceae (Buckthorn Family)

Colubrina

Colubrina asiatica (L.) Brongn.
Kolokolo
Nonopapa, dry limestone flats, 20 ft. alt., *St. John 22,770*. Also observed at Leahi. Rare, but certainly native.

Malvaceae (Mallow Family)

Key to the Genera

Carpels (fruit sections) 1-seeded.
  Style branches stigmatic along inner sides.   *Malva*
  Style branches with terminal stigmas.
    Bracts beneath the flowers 3-1.   *Malvastrum*
    Bracts wanting.   *Sida*
Carpels 2 or more seeded.
  Bracts beneath the flowers wanting; fruit lobed at top.   *Abutil n*
  Bracts present beneath the flowers; carpels fully united.
    Calyx 5-lobed; style 5-lobed.   *Hibiscus*
    Calyx not lobed; style unbranched; seeds hairy.
      Bracts beneath the flowers small, narrow.   *Thespesia*
      Bracts beneath the flowers large, heart-shaped.   *Gossypium*

Abutilon

Abutilon grandifolium (Willd.) Sweet
  A. abutilon (L.) Rusby, and *A. indicum* sensu Forbes, not Don.
Ma'o (=green); *hau hele* (=the *hau* that walks)
Foot of mountain, west side, Jan. 1912, *Stokes*; south half of island, Jan. 1912, *Stokes*; ½ mile east of Kiekie, thicket in gulch, 75 ft. alt., *St. John 22,753*. Also observed at Kii, Kaali Cliff, Kapaka Valley, Mokouia Valley, Keawanui Bay, Haao Valley, Apana Valley, Halulu Lake, Kawaewae and Kamalino. Widely distributed and common. It is a weed, introduced from Peru.

Abutilon incanum (Link) Sweet
 '*Ilima pua kea* (the '*ilima* with white flowers.)
 Foot of plateau, southeast corner, Jan. 1912, *Stokes*; north of Halulu Lake, on dry red dirt, 30 ft. alt., petals white with red color at base, *St. John 23,644*.

Gossypium

Gossypium peruvianum Cav.
 G. barbadense var. *brasiliense (*Macf.) Hutchins. & Stephens
 Pulupulu
 Thicket on dry basalt, Nonopapa, 20 ft. alt., *St. John 22,767*. Introduced from South America, doubtless deliberately for the useful cotton fibers.

Gossypium tomentosum Nutt. ex Seem.
 G. sandvicense Parl.
 Ma'o (= green)
 Headland west of Kaumuhonu Bay, forming thicket on grassy headland, 160 ft. alt., *St. John 22,782*. This was also listed by Forbes (1913: 23) as having been collected by Stokes, but the specimen is not now to be found in the Bishop Museum, nor was it ever entered into the card catalog of the herbarium.

Hibiscus

Hibiscus schizopetalus (Mast.) Hook. f.
 Aloalo; *kula pepeiao*
 Observed in cultivation, probably at Puuwai.

Hibiscus tiliaceus L.
 Paritium tiliaceum (L.) St. Hil.
 Hau
 Observed in a gulch near Halulu Lake. A probable record for this species was published (1789: 85) by Capt. Nathaniel Portlock of the "*King George*" who in June, 1786, visited Yam Bay (Nonopapa). He described the growth near the well half a mile inland as including trees, "about fifteen feet high, and proportionally thick; with

spreading branches, and a smooth bark; the leaves were round, and they bore a kind of nut somewhat resembling a walnut". It was definitely listed by Forbes (1913: 23) as collected in 1912 by *Stokes*, however, the specimen is not now in the Bishop Museum, nor does it appear in the card catalogue of the herbarium as ever having been inserted there. Rare.

Malva

Malva parviflora L.
   Ho-na
   Kii, weed by corral, 15 ft. alt., *St. John 23,582*. Also observed at Kiekie. A weed, introduced from the Mediterranean region.

Malvastrum

Malvastrum coromandelianum (L.) Garcke
   Ha'u-o-i pake (= Chinese vervain)
   Apana, weed in dried muddy bed of old Makanikahau Reservoir, 400 ft. alt., *St. John 23,635*. Also observed at Kii. A weed, introduced, probably from South America.

Sida

Sida fallax Walp.
   'Ilima laukahi (the *'ilima* with a single leaf); *'ilima lau li'i li'i* (*'ilima* with small leaves)
   Kii, Jan. 1912, *Stokes;* Kaali, Jan. 1912 *Stokes*; Nonopapa, in thicket on dry flat, 20 ft. alt., *St. John 22,764*; Kawaewae, dry hillside in brush, 270 ft. alt., *St. John 22,726;* Puu Uala, dry lowlands, 41 ft. alt., March 10, 1977, *C. Christensen 170*. It was also observed at Kalanihale, Kii, Kapaka Valley, Mokouia Valley, Keawanui Bay, Kaununui, Apana Valley, Halulu Lake, Halalii Lake, Kalaalaau Valley and Poooneone. It is an abundant indigenous shrub of the dry lowlands.

Sida rhombifolia L.
   South half of the island, Jan 1912, *J. F. G. Stokes*. A weed introduced from the tropics of the Old World.

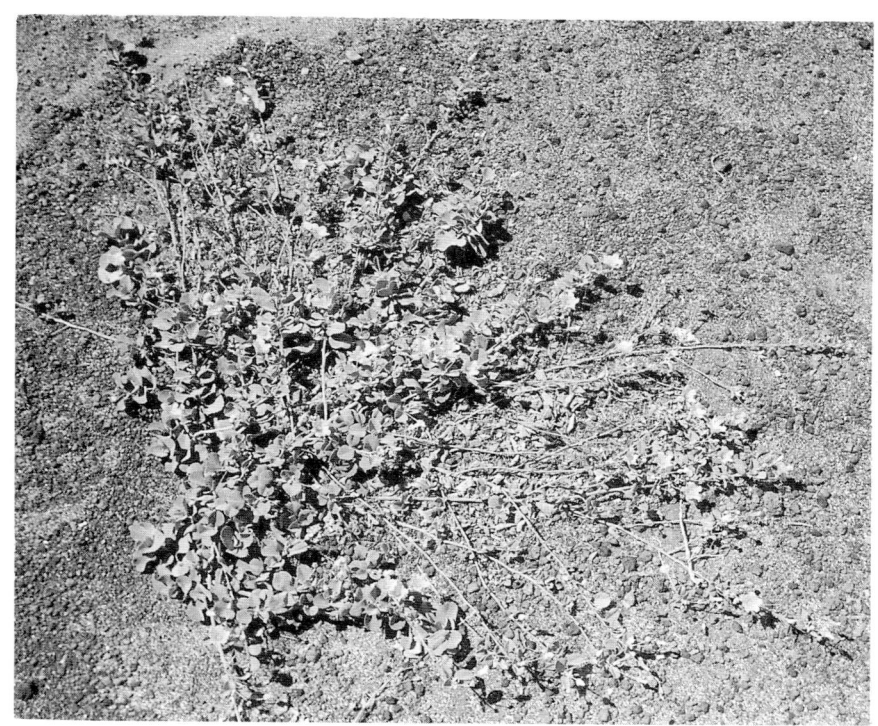

Sida fallax (photo by J.H.R. Plews)

Thespesia

Thespesia populnea (L.) Sol. ex Corrêa
Milo
Foot of plateau, southeast corner, Jan. 1912, *J.F.G. Stokes*; Halalii Lake, north side, edge of lake bed, 20 ft. alt., *St. John 22,748*. Also observed at Halulu Lake and Kalaalaau Valley. Not common. A native of the tropics of the Eastern Hemisphere, introduced to Hawaii by the early Polynesians.

Sterculiaceae (Cocoa Family)

Waltheria

Walterhia indica L.
  W. americana L.
  Hi'aloa; 'uhaloa
  Foot of plateau, southeast, top of talus, 350 ft. alt., Jan. 1912, *Stokes*. Observed at Kii, Kaali Cliff, Kapaka Valley, Poooneone, Kawaewae and Kamalino. A common weedy plant introduced from the tropics of the world.

Tamaricaceae (Tamarix Family)

Tamarix

Tamarix gallica L.
  Paina pupupu (= pine, thick growing); *paina kahakai* (=pine by the beach)
  Kii, 10 ft. alt., planted by house, *St. John 23,563*.

Violaceae (Violet Family)

Isodendrion

Isodendrion remyi St. John, Pacif. Sci. 6: 247-250. *Fig. 12*. 1952.
  Isles Sandwich, Nihau, 1851-55, *J. Rémy 534*. The two sheets of this in the Paris herbarium and the Grey Herbarium are the only ones known. In Hillebrand's Flora (1888: 18) this is incorrectly identified as *I. pyrifolium*, a native of Oahu.

Passifloraceae (Passion Flower Family)

Passiflora

Passiflora foetida L.
  Lani wai (water from heaven)
  Halulu Lake, 50 ft. alt,. vine over bushes, *St. John 22,723*. Also observed at Kalanihale, Kaali Cliff, Kapaka Valley, Mokouia Valley,

Keawanui Bay, Puuwai, Haao Valley, Apana Valley, Kiekie, Nonopapa, Halalii Lake, Kawaewae, Kamalino, Leahi and Kawaihoa Point. It is very common, especially in the dry lowlands, and its bright red fruits are a favorite food of the wild turkeys and peacocks. Its vernacular name, used only on Niihau, would seem to indicate that the people also appreciated it. They likened the small globules of watery pulp around the few seeds to a thirst-slaking gift of the gods. It is a weed introduced from Brazil.

Caricaceae (Papaya Family)

Carica

Carica papaya L.
　He'i
　Observed at Mokouia Valley, Keawanui Bay, Puuwai and Kiekie. This is a cultivated fruit tree, introduced from tropical America.

Cactaceae (Cactus Family)

Cereus

Cereus peruvianus (L.) Mill.
　Observed in cultivation at Kiekie. It was introduced from tropical South America.

Hylocereus undatus (Haw.) Britt. & Rose
　Observed in cultivation at Puuwai. It was introduced from Mexico.

Opuntia

O. megacantha Salm-Dyck
　O. tuna, sensu Forbes, not of Mill.
　Pa pipi (= fence for the cattle); less frequently *pa nini*
　Observed at Kaali Cliff, Kapaka Valley, Mokouia Valley, Pueo Pali, Keawanui Bay, Puuwai, Haao Valley, Apana Valley, Kaeo, Halulu Lake, Halalii Lake, Kawaewae, Kamalino, Leahi and Kawaihoa Point. The writer took no specimens, but has six

photographs of it. Forbes recorded it (1913: 23) but there is no trace that any of Stokes' specimens were preserved. This plant was widely distributed and abundant from near the seashore to the top of the mountain. Stearns reported (1949: 6) that it, "formerly the chief food for animals during a drought, has been rapidly dying out because of an accidentally introduced disease or insect, not yet identified". The disease was first noticed on Kauai about the year 1943, then it soon appeared on Niihau. It attacked the cactus like a wild-fire blight, whole plants turning yellow, then gray with black pustules, then the whole sturdy plant collapsing in a pile of soft, rotting tissue. Some joints may sprout new shoots and live for a while, but soon they are stricken and die. Officials of the Board of Agriculture and Forestry have stated the disease was caused by fungus, a species of *Fusarium*. In many regions a blight that would kill out the cactus would be a blessing, but not so on Niihau. On this arid island, the cactus was the principal source of water for grazing animals. As droughts of six months or more are common, the abundant cactus was a great resource. Its near destruction reduced the carrying capacity of the range by 600 head of cattle. The cactus blight was a disaster to Niihau.

Punicaceae (Pomegranate Family)

Punica

Punica granatum L.
 Pomelaiki (= a Hawaiianization of pomegranate)
 Observed at Puuwai and Apana Valley. It is a cultivated fruit tree introduced from the Mediterranean region.

Myrtaceae (Myrtle Family)

Key to the Genera

Fruit fleshy.
 Calyx 4-lobed, not splitting in flowering; seeds 1-2, large.
 Eugenia
 Calyx not lobed, but splitting in flowering; seeds many.
 Psidium
Fruit a dry capsule. *Eucalyptus*

Eucalyptus

Inflorescence many-flowered; fruit 10 mm. wide; herbage lemon-scented.
                  E. citriodora
Inflorescence few-flowered.
Flower stalks flattened; flowers 4-12; fruit 10-12 mm. wide.
                   E. robusta
Flower stalks rounded; flowers 4-8; fruits 6 mm. wide.
                 E. camaldulensis

Eucalyptus camaldulensis Dehnh.
 Pale-piwa (= to keep off the fever)
 Apana Valley, 500 ft. alt., *St. John 22,796*. A cultivated tree introduced from Australia.

Eucalyptus citriodora Hook.
 Paniau, 900 ft. alt., planted tree, lacking the lemon-scented oil, *St. John 23,587*. The writer was unable to determine this specimen, so sent it to Australia. Dr. R. M. Anderson of the Sydney Botanic Gardens reported the above determination.

Eucalyptus robusta Sm.
 Pale-piwa (= to keep off the fever)
 Apana Valley, planted tree, 500 ft. alt., *St. John 22,795*. It was introduced from Australia.

Eugenia

Eugenia cumini (L.) Druce
 E. jambolana Lam.
 Palama (= Hawaiianization of plum)
 Observed in cultivation at Mokouia Valley, Puuwai, and Apana Valley. It is a fruit tree introduced from tropical Asia.

Eugenia uniflora L.
 Observed in cultivation at Kiekie. It is a fruit bearing shrub introduced from Brazil

Psidium

Fruits yellow, about 5 cm. in diameter; leaves strongly veiny; branchlets 4-angled. *Psidium guajava*
Fruits purplish-red, about 25 mm. in diameter; leaves inconspicuously veiny; branchlets not angled.
P. cattleianum var. *littorale*

Psidium cattleianum var. *littorale* (Raddi) Fosb.
  Waiawi
  Observed at Halulu Lake. Introduced from America for its scarcely edible fruits.

Psidium guajava L.
  Puawa (= a Hawaiianization of guava)
  Observed in cultivation at Puuwai. It is a fruit tree introduced from tropical America.

Araliaceae (Ginseng Family)

Cheirodendron

Cheirodendron trigynum (Gaud.) A. Heller var. *hillebrandii* Sherff
  Panax ovatum Hook. Arn., Bot. Beechey Voy. 84. 1832.
  Not found by any recent collector, the original record being from a collection of a leaf specimen by Lay and Collie of the Beechey Voyage. It is certainly extinct now. The type specimen was studied by the writer in 1954. It is a sterile one, but it is unlike the known varieties of the species on Kauai. It does match the foliage of any on Oahu, so it is tentatively identified as var. *hillebrandii*. It was described and illustrated by St. John (1959: 171-172, fig.6).

Reynoldsia

Reynoldsia sandwicensis A. Gray
  'Ohe'ohe
  The only record of this is in the publication by C. N. Forbes (1913:23) based on the collection of Stokes. The species is there listed

without comment. The specimen is not now in Bishop Museum, nor does it appear in the card catalog of the herbarium as ever having been inserted there. When exploring Niihau in 1947, the writer did not see this tree. It was one of the four native trees reported by Forbes, so its rediscovery was much desired. When asked about it, the guide K. Niau, immediately replied that he knew the ʻoheʻohe, and that it was now very rare. He said there was only one living tree left, and then led the way on Aug. 13 to a spot on a steep side of upper Kanaha Valley. At the site which he knew, we found no sign of a tree, it having died, and the soft stems having rotted away since the guide's last visit to the spot. Without doubt it is now extinct on Niihau. The record for Niihau is a reasonable one, as the genus once, and to some degree still, makes a scattered arborescent growth on the low, very dry sections of all the Hawaiian Islands. In the revision of the genus by Sherff (1952: 7), he states "I have seen as yet no specimens from Kauai, Niihau, or Kahoolawe". Degener and Degener (Fl. Haw. 281: VIII/5/68) mention the lack of any voucher specimen for Forbes' record.

Umbelliferae (Carrot Family)

Hydrocotyle

Hydrocotyle verticillata Thunb.
Lau-kahi (= leaf, to rub)
Kaali, Jan. 1912,*Stokes*; Kaali Cliff, common in wet muck on ledge by Kaali Spring, 750 ft. alt., *St. John 22,834*.

Primulaceae (Primrose Family)

Lysimachia

Lysimachia mauritiana Lam.
L. spathulata (Vent.) Klatt
This was credited by Hillebrand (188;285) from "Maui and Niihau" (Rémy); and Rémy's number list at the Museum National d'Histoire Naturelle, Paris has Rémy 460, "Molokai, Kauai, ou Nihau". The specimens of this collection in the Gray Herbarium have

the data, Rémy 460, Nihau, while no. 459 is marked from Maui. This Indo-Pacific littoral species is known to occur in the Hawaiian Islands, on windward Kauai, Molokai and Maui.

Plumbaginaceae (Leadwort Family)

Plumbago

Plumbago zeylanica L.
   Lauhihi (= leaf; to entwine)
   Kaali, Jan. 1912, *Stokes*; Poooneone, sand dunes, 15 ft. alt., *St. John 22,741*; Puuwai, dry plains, 10 ft. alt., Feb. 12, 1977, *C. Christensen, 141*. Also observed at Kii, Kaali Cliff, Mokouia Valley, Paniau, Puuwai, Halalii Lake, Kamalino and Leahi. Occasional from the seashore to the mountain top, but less frequent in the hills. A native shrub.

Sapotaceae (Sapodilla Family)

Key to the Genera

Sepals 5; stamens 5.   *Chrysophyllum*
Sepals 6 in 2 whorls; stamens 6; staminodia 6.   *Manilkara*

Chrysophyllum

Chrysophyllum oliviforme L.
   Palama waiu
   One mile N. of Leahi, Aug. 13, 1947. *St. John 22,775*. According to Mrs. A. Robinson, this was introduced about 1902. Forbes reported *C. polynesicum* (= *Nesoluma polynesicum*), (1913: 24), based upon a collection or report by Stokes. There is no such specimen in the Bishop Museum now, nor is it catalogued as having been accessioned there. It is likely to have been *C. oliviforme*. The black fruits are edible. It grows vigorously from the moist bottom of the sink hole, and its green, flourishing crown expands. It is a cultivated fruit tree introduced from the West Indies.

Manilkara

Manilkara zapota (L.) van Royen
Achras zapota L.
Sapota achras Mill.
Manilkara zapotilla (Jacq.) Gilly
Momona (= sweet)
Puuwai, in gardens, *St. John 22,810*. A cultivated fruit tree introduced from South America.

Oleaceae (Olive Family)

Key to the Genera

Leaves compound; corollas 4-9 lobed. *Jasminum*
Leaves simple; corollas 4-lobed.
  Leaves broad, notched at the tip; stamens attached at base of corolla; flowers yellow. *Noronhia*
  Leaves lance-shaped; stamens attached to the corolla; flowers white. *Olea*

Jasminum

Jasminum sambac (L.) Ait.
Observed in cultivation at Puuwai. An ornamental shrub introduced from India.

Noronhia

Noronhia emarginata Stadm.
Palama waiu (= plum, milk)
Kiekie, by house, 50 ft. alt., *St. John 22,788*. A cultivated fruit tree introduced from Madagascar.

Olea

Olea europaea L.
Oliwa haole (= white man's olive)
Observed in cultivation at Apana Valley and Kiekie. A fruit tree introduced from the Mediterranean.

Apocynaceae (Periwinkle Family)

Catharanthus

Catharanthus roseus (L.) G. Don f. *albus* (Sweet) G. Don
Vinca rosea var. *alba* Sweet.
Observed in cultivation and also escaped in the village, Puuwai. Introduced from tropical America.

Asclepiadaceae (Milkweed Family)

Key to the Genera

Herbs; flowers red and yellow. *Asclepias*
Shrubs; flowers white or lavender. *Calotropis*

Asclepias

Asclepias curassavica L.
Nuʻumela (= a numeral)
South half of island, Jan. 1912, *J.F.G. Stokes*; ridge south of Haao Valley, weed in *Opuntia* pasture, 350 ft. alt., *St. John 23,641*; Poooneone, thicket near beach, 15 ft. alt., *St. John 22,742*. A weed introduced from tropical America.

Calotropis

Calotropis gigantea (L.) R. Br. ex Ait. f.
Liliʻu (= a part of the name of Liliuokalani, a Hawaiian queen)
Observed in cultivation by the residence at Kiekie. An ornamental introduced from India.

Convolvulaceae

Key to the Genera

Leafless parasite. *Cuscuta*
Green, nonparasitic plants.
  Corolla tiny, 5-lobed, white, enclosed by the calyx; a prostrate herb. *Cressa*
  Corolla unlobed, large, trumpet-like, protruded.
    Stigmas 2, club-shaped; prostrate herb with small blue flowers. *Jacquemontia*
    Stigmas rounded.
      Stigmas 2; leaf blades palmately divided, long brown hairy. *Merremia*
      Stigmas 1, leaf blades variable, not long brown hairy. *Ipomoea*

Cressa

Cressa insularis House
  Puuwai, mud flat, Feb. 12, 1977, *C. Christensen 150;* March 17, 1977, *Christensen 171*. Native to the shores of Hawaii and California.

Cuscuta

Cuscuta sandwichiana Choisy
  Kauno'a-lei
  Kamalino, on beach, 10 ft. alt., parasite on *Ipomoea brasiliensis, St. John 22,774*. Also observed near Puuwai. Native only to the Hawaiian Islands.

Ipomoea

Blades palmately cut to the base. *I. cairica*
Blades not so.
  Terrestrial creepers, rooting at joints.
    Blades heart-shaped at base. *I. batatas*
    Blades rounded at base.
      Corolla white; stems buried in beach sand. *I. stolonifera*
      Corolla pink; stems on the surface. *I. brasiliensis*

Climbing vine; blades ovate heart-shaped; corolla blue, fading pink. *I. indica*

Ipomoea batatas (L.) Lam.
'Uala

Observed by the writer in the village of Puuwai, where it was cultivated by the native Hawaiians in their gardens.

Previously anthropologists had collected it there. Stokes obtained three good collections, with ample herbage, and two of the three with flowers and fruit. They were all labelled, "South half of Island". No vernacular varietal names were recorded. The first Stokes collection has blades 4-6.5 cm. long, 3-5 lobed, ¾ way into oblong, acute lobes. Handy later annotated them, "Looks like *wai ani-ani* from Molokai," and the writer would add, also much like the variety is collected on Niihau by Handy as "*'uala 'ele'ele o kohala*". The second has blades 4-7 cm. long, 5-lobed ¾ way to base, the central lobe oblong, the lateral ones narrowly, linear-lanceolate. It is unique among the Niihau collection, but Handy has annotated the plant as "like *kala* from Maui". The third collection has 6-7.5 cm. long deltoid-cordate, acuminate entire leaves. This is unique among the Niihau collection, but Handy annotated it, "looks like *hua moa...*" He illustrates this (1940: 134, fig.13e), but the writer observed that this Niihau plant had blades differing by being more deltoid, and tapering to a longer, acuminate apex. It appears to tally much more closely with the variety *apo*, as illustrated by Handy (1940; fig. 13b).

The following eight collections with vernacular names were obtained on Aug. 14 by E. S. C. Handy. They all consist of detached leaves, without stems, flowers, or fruits. The first is *uala palani*, with blades 7.5-9 cm. long, broadly rounded cordate, sub-acuminate, usually with 1-2 dentations at the widest part. The varietal name is listed by Handy (1940: 142) but not illustrated or described. It is probably identical with his variety "*kihikihi poepoe*" (1940: fig. 14c). The second is *uala kalia*, with blades 10.5 cm. long, deltoid, acuminate, entire, except for the usually subacuminate tips to the lateral points of the triangle. Handy list this variety (1940: 142), but does not describe or illustrate it. The third is *'uala Kamalino*, with blades 6-8 cm. long, 7-10.5 cm. broad, broadly deltoid, usually 5-lobed ½ way to base, the central lobes oblong or broadly deltoid, sub-acuminate, the deltoid lobes confluent into a large quadrate base. The Hawaiian word *Kamalino* means Samuel's. The variety listed and illustrated by Handy (1940: 142, fig. 13i). The fourth is *'uala Molokai*, with blades 7-8 cm. long, 7.3-8.2 cm.wide, 3-lobed, the central lobe narrowly deltoid, acute, lobed ½ way, the base transversely oblong, truncate, the lateral lobe tips slender lanceolate, obtuse, ascending. It is listed and illustrated by Handy (1940: 142, fig. 15d).

The fifth variety is *uala eleele o Kohala*, with blades 5-6 cm. long, 6.5-10 cm. wide, palmately 3-5 parted ⅘ way to base, the lobes narrowly lanceolate, wide-spreading. This is listed by Handy as "*'ele'ele Kohala*",(1940: 142). The sixth is *uala Wailua*, with blades 6-7 cm. long, sub-orbicular in outline of the tips, the base sub-truncate, palmately 5-parted ⅘ way to base, the central lobe elliptic, acute, the lateral lobes narrowly deltoid, tapering, at base confluent. This is listed and figured by Handy (1940: 143, fig. 16h). The seventh is *uala Papa'a kowahi*, with blades 7.5-11 cm. long, 10-12 cm. wide, palmately 5-parted ¾ way to the broad sub-truncate base, the central lobe lance-elliptic,the lateral lobes narrowly deltoid-lingulate, obtuse, sub-apiculate. It is listed and figured by Handy (1940: 142, fig. 15k). The eighth is *uala manamana = piko nui*, with blades 6.8-7 cm. long, 8.5-10.5 cm. wide, otherwise like the preceding. Handy figures the latter (1940: 142, fig. 16c).

Previously the *uala*, or sweet potato, was the staple crop of the natives on Niihau, as Handy states (1940: 152-153). He quotes from the manuscript of Samwell (in the Archives of Hawaii) who visited the island with Capt. Cook, "Neehau for the most part consists of low land entirely bare of trees. The soil is rich and capable of producing all kinds of fruit, was it properly cultivated, but as the island is thin of inhabitants, the small patches which are here and there planted with yams and sweet potatoes afford a sufficient supply for them, while large plains of fine land is suffered to waste...saw a few plantation of sugar cane and plantain and two or three palm trees...We procured yams enough here to serve the ships for bread for six weeks...The natives cultivate more sweet potatoes than yams."

Ipomoea brasiliensis (L.) Sweet
  I. pes-caprae (L.) Sweet, sensu Hawaiian authors.
Pohuehue (= rounded)
Foot of plateau, southeast, Jan. 1912, *J. F. G. Stokes*. Observer by the writer as common on coral sand beaches, at Kalanihale, Kii, Kapaka Valley, Kaununui, Puuwai, Kiekie, Nonopapa, Kamalino, Leahi and Kawaihoa Point. It is native to Hawaii and the world tropics.

Ipomoea cairica (L.) Sweet
  I. tuberculata var. *trichosperma* Hbd.
Koali 'ai'ai (= the edible koali)

Ipomoea brasilliensis (photo by J.H.R. Plews)

Foot of mountain on west side, Jan. 1912, *J. F. G. Stokes*; Kaali Cliff, 59 ft. alt., vine over sunny rock knoll, flowers pink-magenta, *St. John 23,580*; Kamalino, 50 ft. alt., climbing over stones, *St. John 23,627*. Observed also at Haao Valley.

Ipomoea indica (J. Burm.) Merr.
   I. insularis Steud.
   Koali la'au (= medicinal morning glory)
   Foot of mountain, on west side, Jan. 1912, *J. F. G. Stokes*, two sheets; Kamalino, 20 ft. alt., vine in thicket, flowers pale bluish, *St. John 22,773*. Also observed at Kii and Paniau.

Ipomoea indica (J. Burm.) Merr. f. *alba*, forma nova
   Koali (= morning glory)
   One mile northeast of Leahi, 70 ft. alt., vine over trees from sinkhole in arid limestone flat, flowers white, Aug. 13, 1947, *St. John 22,776* (Holotype, Bish.).
   Floribus albis: Though this is a wide-ranging species, and variant with albino corollas are to be expected, no description of such a one has been found.

Ipomoea stolonifera (Cyrill.) J. F. Gmel.
   I. acetosaefolia (Vahl) R. & S.
   I. littoralis (L.) Boiss. (1879), not of Blume (1826).
   Convolvulus stolonifera Cyrill.
   Hunakai (= foam of the sea)
   Niihau, 1851-1855, *J. Rémy*. This specimen, doubtless seen in Paris or at the Gray Herbarium, was listed by Hillebrand (1888: 314). It is also Listed as *Rémy 416* from Niihau, in the manuscript number list of Rémy plants in the Museum National d'Histoire Naturelle, Paris.
   Kaununui, 10 ft. alt., trailing, nearly buried in coral sand of foredune, *St.John 22,717*.
   This native species still exists on several of the Hawaiian Islands, but it is not common anywhere. The corollas have the lower part of the tube pale greenish outside, but inside the eye is greenish-yellow with the remainder all snow white. They look like white circles on the coral sand, and well deserve the poetic name "hunakai" (foam of the sea).

Jacquemontia

Jacquemontia sandwicensis A. Gray var. *sandwicensis*
  Pa'u a Hi'iaka (= the skirt of Hi'iaka, sister of Pele)
  Kii, Jan. 1912, *J. F. G. Stokes*; Kawaihoa Point, 100 ft. alt., trailing on coral sands, *St. John 23,618*. Also observed at Kalanihale, Kii, Kaununui, Nonopapa, Halulu lake, Poooneone, Kawaewae, Kamalino, Leahi and Kawaihoa Point. Abundant just above the beaches, but also sporadic over the lowland flats, and on the mountain even to the very summit.

Merremia

Merremia aegyptia (L.) Urb.
  Ipomoea pentaphylla (L.) Jacq.
  Recorded by Forbes (1913: 24) as one of the collections of J. F. G. Stokes, but there is no such specimen now in the Bishop Museum, nor is it in the card catalogue as ever having been entered in the herbarium. However, it is an abundant and easily recognized lowland weed, and there seems no reason to doubt this record by Forbes. The species is an introduction from the Old World Tropics.

Hydrophyllaceae   (Water-leaf Family)

Nama

Nama sandwicensis A. Gray
  Hinahina kahakai (= silvery on the beaches)
  On partly stabilized sand dune back of foredune, Kalanihale, 15 ft. alt., *St. John 22,711*; sand dunes opposite mouth of Kanalo Valley, Dec. 30, 1947, *A.E. Robinson*; coral sand flats back from beach, Poooneone, 15 ft. alt., *St. John 22,743*. Also observed at Kawaihoa Point. This species is listed by Forbes (1913: 24) as collected by J.F.G. Stokes. However, the specimen, so labeled in Forbes' hand is a flowering specimen of *Lipochaeta integrifolia*. It seems incredible that Forbes could have made such a blunder, and seems likely that there was a confusion in handling, and in interchange of labels. There is, however, no *Nama* specimen labeled *Lipochaeta*.

Boraginaceae (Heliotrope Family)

Key to the Genera

Fruits splitting into 4 nutlets; shrubs or herbs. *Heliotropium*
Fruits not splitting into 4 nutlets; trees.
  Style 2-parted. *Cordia*
  Style simple; tree with hairy leaves. *Messerschmidia*

Cordia

Fruit white, edible. *C. sebestena*
Fruit green to yellow, hard. *C. subcordata*

Cordia sebestena L.
  Kou haole (= the white man's kou)
  Half mile east of Kiekie, 75 ft. alt., in shaded gulch, planted tree, *St. John 22,754*. An ornamental tree introduced from tropical America.

Cordia subcordata Lam.
  Kou
  Cultivated in hedge by house, Kiekie, 50 ft. alt., *St. John 22,789*; mouth of Kaumuhonu Valley, 50 ft. alt., dry flat, with *Prosopis, St. John 23,630*. Source of valuable wood to the native Hawaiian on the other islands. There is no early record of it on Niihau. Introduced from southern Asia.

Heliotropium

Heliotropium anomalum var. *argenteum* A. Gray
  Nohonoho pu'uone (= to sit around; sand dunes)
  Kaali, Jan. 1912, *J.G.F. Stokes*. Sand hills opposite mouth of Kanalo Valley, Dec. 30, 1947, *A.F. Robinson*; on front dune, in coral sand, Kaununui, 15 ft. alt., *St. John 22,714*. Also observed at Kalanihale, Kii, Kamalino and Kawaihoa Point. It is a native plant characteristic of the coral sand dunes near the shore.

Heliotropium curassavicum L.
Po'opo'ohina (=gray haired); *lau po'ohina* (=leaf, gray)
Sand dunes opposite mouth of Kanalo Valley, Dec. 30, 1947, *A.F. Robinson*; south border, Halalii Lake, exsiccated salty flat, 30 ft. alt., *St. John 22,736*; ponds, southern end, Jan. 1912, *J.F.G. Stokes*; Kawaihoa Point, north base, wet seep on cliff, *St. John & L.D. Tuthill 22, 786*. Also observed at Kalanihale, Poooneone and Kamalino. It is common on saline spots.

Messerschmidia

Messerschmidia argentea (L.f.) I.M. Johnston
One mile N. of Kamalino, top of beach, 5 ft. alt., two young volunteers from sea drift, *St. John 23,610*. Also observed on the beach, two miles north of Poooneone. It has been introduced from the tropical Indo-Pacific.

Verbenaceae (Verbena Family)

Key to the Genera

Flowers without stalks (pedicels).
  Fruit of 4 nutlets; herb.   *Verbena*
  Fruit fleshy, with 2 seeds; prickly shrub.   *Lantana*
Flowers all with stalks (pedicels); flowers blue.
  Berries yellow; leaves hairless.   *Duranta*
  Fruits brown, hard; leaf white hairy beneath.   *Vitex*

Duranta

Duranta repens L.
Pa Kanalo Iki, dry scrubland, 1,000 ft. alt., Jan 15, 1977, *C. Christensen 114*. This is an ornamental shrub, introduced from tropical America.

Lantana

Lantana camara L.
Lanakana (= a Hawaiianization of lantana)
One mile northeast of Leahi, in limestone sinkhole, in shade of *Artocarpus*, 70 ft. alt., rare, *St. John 22,779;* Kawaihoa Point, on grassy slopes, 200 ft. alt., *St. John 23,613*. It was also observed at Paniau, but it is rare. Stearns (1947: 6) reported that this thorny shrub is "seeded by birds from Kauai, but is removed whenever it is found".

Verbena

Verbena litoralis Kunth
V. bonariensis sensu C.N. Forbes (1913: 24), not of L.
Ha'u-o-i
Makanikahau Reservoir, Apana, common on dried mud, 400 ft. alt., *St. John 23,633;* south half of island, Jan. 1912, *J.F.G. Stokes;* also ponds, southern end, *Stokes*. Also observed at Makouia Valley, Halulu Lake, Halalii Lake and Poooneone. It is a weed introduced from tropical America.

Vitex

Vitex ovata Thunb.
V. trifolia var. *simplicifolia* Cham.
Kolokolo (= to crawl)
On coral sand dunes by beach, Kaununui, 10 ft. alt., *St. John 22,718*; Kiekie, Jan. 1912, *J.F.G. Stokes*. Also observed at Kii, Puuwai, Poooneone, Kamalino and Kawaihoa Point. It is a trailing, or ascending native shrub of the coastal sand dunes.

Labiatae (Lamiaceae) **(Mint Family)**

Key to the Genera

Stamens 2; flowers blue. *Salvia*
Stamens 4.
    Calyx in fruit ¼ in. long; leaves sharp pointed; flowers white to bluish. *Ocimum*

Calyx in fruit ⅛ in. long; leaves blunt; flowers blue.
Plectranthus

Ocimum

Ocimum basilicum L.
Ki hohono (= tea, strong smelling)
Half mile northeast of Kiekie, patch by roadside, 20 ft. alt., *St. John 22,792*; foot of plateau, southeast, Jan. 1912, *J.F.G. Stokes*, but this specimen was not included in the enumeration by C.N. Forbes. It was observed at Leahi. It is a flavoring herb introduced from western Asia.

Plectranthus

Plectranthus parviflorus Willd.
P. australis R. Br.
Oneeheow (=Niihau), *Lay & Collie*, in 1826; reported by Hooker and Arnott (1832: 92); also Jan. 1912, *J.F.G. Stokes*. Observed at Kaali Cliff, Paniau, Apana Valley and Leahi. It is a weed introduced from Australia.

Salvia

Salvia occidentalis Sw.
Priva aspera sensu Forbes (1913: 24), not of Kunth
Kaali, Jan. 1912, *J.F.G. Stokes;* on rocky cliff, Pueo Point, 1,200 ft. alt., *St. John 22,804*. It was also observed at Mokouia Valley. It is a weed introduced from tropical America.

Solanaceae (**Nightshade Family**)

Key to the Genera

Fruit a dry pod; herbs; corolla lobes widespreading. *Nicotiana*
Fruit a berry.
   Flowers solitary.
      Corolla lobes widespreading; fruit bright-colored; leaves with many veins. *Capsicum*

Corolla lobes ascending; berry red; only the midrib visible in the leaves. *Lycium*
Flowers in clusters.
Anthers fertile throughout, opening at top by slits or pores.
Solanum
Anthers with narrowed sterile tip, opening lengthwise down the sides. *Lycopersicon*

Capsicum

Capsicum annum L.
Nioi
Observed at Apana Valley and Kawaewae. It is both cultivated and established. This is a common chile pepper, a native of tropical America.

Lycium

Lycium sandwicensis A. Gray
L. *carolinianum* var. *sandwicense* (A. Gray) C. L. Hitchc.
'Ae'ae
On coral ledge on beach, Kaununui, 5 ft. alt., *St. John 22, 715*; south corner of plateau, Jan. 1912, *J.F.G. Stokes*. Also observed by the writer at Kii, Kalaalaau Valley, Poooneone, Kamalino, Leahi and Kawaihoa Point. This shrub is native to Hawaii and to Polynesia.

Lycopersicon

Lycopersicon esculentum Mill.
'Ohi'a
Kaali, Jan. 1912, *J.F.G. Stokes*, identified as the species; Kaali Cliff, wet soil on ledge by Kaali Spring, 750 ft. alt., *St. John 22,833*. Introduced from South America.

Nicotiana tabacum L.
 Paka
 Kaali, Jan. 1912, *J.F.G. Stokes*; half mile northeast of Kiekie, rocky slope, 75 ft. alt., *St. John 22,752*. Introduced from South America.

Solanum

Shrub; leaves softly hairy; corolla hairy; anthers narrowed to tips.
 S. *nelsonii*
Herb, without hairs; anthers broad at the tip. S. *nigrum*

Solanum nelsonii Dunal in A. DC.
 Akia (=bitten by)
 Kawaewae, under *Prosopis* thicket, low shrub, 75 ft. alt., *St. John 22,731*; Leahi, in coral sand near beach, depressed shrub, *St. John 23,621*; south end, sand country, Nov. 1, 1939, *G.C. Munro*. Occasional on flats near the shore. The species is endemic to the Hawaiian Islands.

Solanum nigrum L.
 S. nodiflorum Jacq.
 Popolo
 Kaali, Jan. 1912, *J.F.G. Stokes*; Nonopapa, rocky edge of dried, salty inlet, 10 ft. alt., *St. John 22,763*. Also observed by the writer at Kaali and Kawaewae.

Scrophulariaceae (**Figwort Family**)

Bacopa

Bacopa monnieri (L.) Wettst.
 'Ae'ae
 Kaali, Jan. 1912, *J.F.G. Stokes*, this is the basis for C.N. Forbes' published record (1913: 21) of *Sesuvium portulacastrum;* in wet soil

in overflow of Kaali Spring, 750 ft. alt., *St. John 22,832*. It is suggested that this was introduced with taro from Kauai. The plant is native to Hawaii, and to many other tropical and temperate lands.

Bignoniaceae (Bignonia Family)

Tecomaria

Tecomaria capensis (Thunb.) Spach
Makahala
Observed in cultivation near the residence at Kiekie. This is an ornamental introduced from South Africa.

Myoporaceae (Naio Family)

Myoporum

Myoporum sandwicense A. Gray var. *sandwicense*
Niihau form, see Webster, Pacif. Sci. 5: 63, 1951.
Naio
Mokouia Valley, 500 ft. alt., rocky ledge in *Prosopis* grove, *St. John 23,592*. One mile south of Paniau, crest of windward cliff, 1,200 ft. alt., *St. John 23,603;* steep rocky shore east of Kalaalaau Valley, basalt talus above beach, 50 ft. alt., *St. John 22,747*; south corner of plateau, Jan. 1912, *J.F.G. Stokes*.

Rubiaceae (Coffee Family)

Morinda

Morinda citrifolia L.
Noni
Talus, north Kona cliffs, Jan. 1912, *J.F.G. Stokes;* near Halulu Lake, thicket in gulch, 50 ft. alt., *St. John 22,722*. Also observed at Mokouia Valley, Keawanui Bay, Apana Valley, Halalii Lake and Kamalino. An aboriginal introduction from southeastern Asia.

Cucurbitaceae (Gourd Family, Squash Family)

Key to the Genera

Fruit dry, hairy, 1-seeded. *Sicyos*
Fruit fleshy, many seeded.
  Fruit orange, splitting open. *Momordica*
  Fruit not splitting.
    Corolla 5-lobed halfway. *Cucurbita*
    Corolla 5-lobed almost to the base.
      Tendrils branched; fruit smooth. *Citrullus*
      Tendrils simple; fruit densely bristly. *Cucumis*

Citrullus

Citrullus lanatus (Thunb.) Matsu. & Nakai
  C. vulgaris Schrad.

This watermelon is occasionally cultivated now, and there is an early record by Capt. Broughton (1804: 46) of fruits being grown and traded by the natives in 1796.

Cucumis

Cucumis dipsaceus Ehrenb. ex Spach
  North Kona Cliff, talus, Jan. 1912, *J.F.G. Stokes*. Observed by the writer at Leahi. A weed introduced from tropical America.

Cucurbita

Cucurbita pepo L.

The pumpkin was apparently early cultivated, for on Feb. 19, 1796, Capt. William R. Broughton reported (1804: 46) trading for "yams, potatoes, watermelons and pumpkins". The seeds were introduced from America.

Momordica

Momordica charantia L.
'Ai-a-ka-manu (= food of the bird); *palea*
One mile south of Paniau, weed in grassy range land, 1,150 ft. alt., *St. John 23,604;* Nonopapa, trailing in grass, 20 ft. alt., *St. John 22,760.* Also observed at the village Puuwai. It was introduced from tropical Asia and is here a weed.

Sicyos

Sicyos niihauensis St. John, 1959.
Pua-o-kama
One mile west of Kii, 50 ft. alt. (also common on Kaali cliff), grassy, sandy flats, elongated vine climbing *Prosopis* tree, *St. John* 23,567 (Holotype, BISH).

Lobeliaceae **(Lobelia Family)**

Stem thick, fleshy; corolla white or yellow, salver-shaped, symmetrical. *Brighamia*
Stem woody.
Inflorescence terminal; fruit a dry capsule. *Lobelia*
Inflorescence from the leaf axis; fruit a berry. *Delissea*

Brighamia

Brighamia insignis A. Gray in Mann
'Olulo, pu aupaka (= shell used as a trumpet, the old stem becoming hollow)
This monotypic genus was described by Gray from a collection by *W.T. Brigham* on Molokai and from *Rémy 309 ter* from Kauai or Niihau. Hillebrand (1888: 235) attributed it to Niihau, *Rémy,* but neither the holotype at the Gray Herbarium or the number list at the Paris herbarium indicates any choice between the alternative localities. It is now known that *Brighamia* grows on both islands. In his monograph of the genus of three species and two varieties, the present writer (1969) made the Rémy collection in the Gray

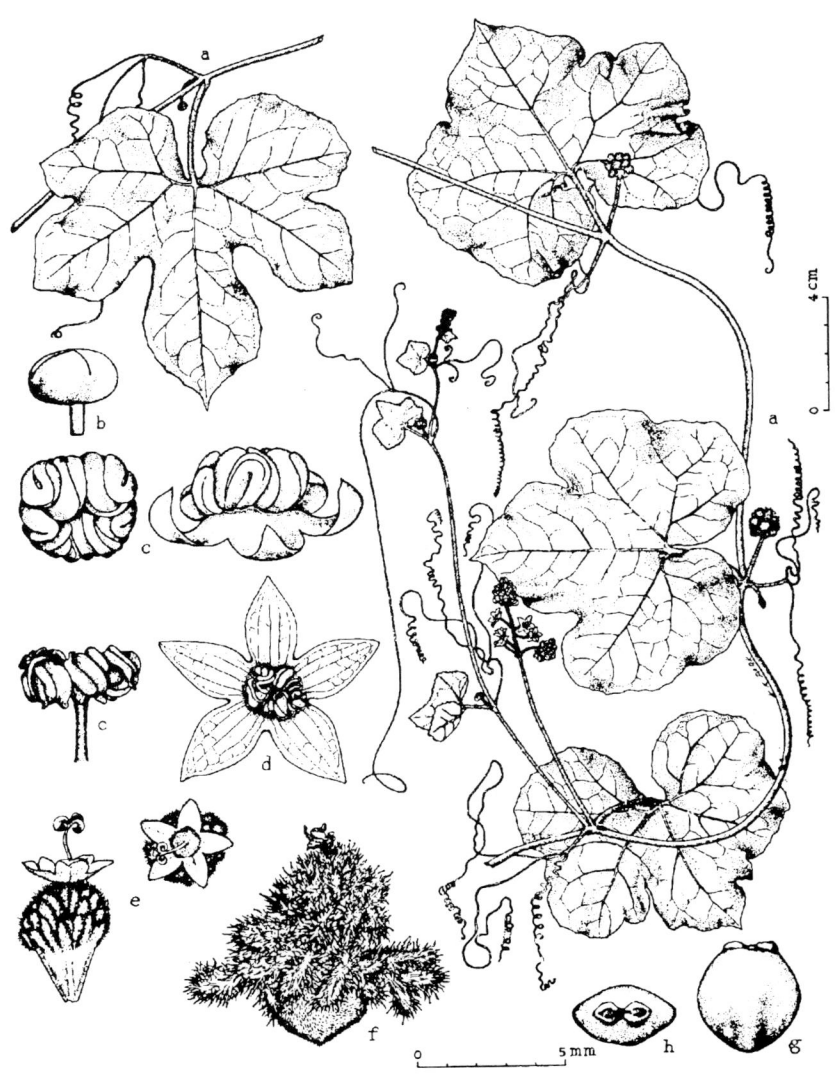

Sicyos niihauensis, from holotype: *a*, habit X 1/2; *b*, staminate bud X 5; *c*, anthers X 10; *d*, staminate flower X 5; *e*, pistillate flower X 5; *f*, fruit X 5; *g*, seed, lateral view, X 5; *h*, seed, apical view, X 5.

Herbarium the lectotype of the species *B. insignis*. He also proved that it differed from the two kinds on Kauai by having the filament tube glabrous and the seed coat covered by conic tubercles. From that it was deduced that the locality for *B. insignis*, the type species of the genus, was Niihau.

For Niihau it was recorded by Forbes (1913: 25), "Mr. Stokes observed on inaccessible cliffs". In 1947 the writer also observed it on cliffs above Kaali Spring. He was dissuaded by his guide from attempting to reach it. It is reported that the flowers are yellow. The trunks grow to heights of from 3 to 8 feet.

Delissea

Delissea niihauensis St. John, Pacif. Sci. 13: 177, 179-181, fig. 8, 1959.

The holotypic specimen is Niihau, *W.T. Brigham*. Also, there is Nihau, 1851-1855, *J. Rémy 300 bis*. The plant is a shrub, endemic to Niihau, with ovate, pointed leaves, and clusters of tubular flowers 22-25 mm. long. Doubtless it had milky sap.

Brigham said "this is the only *Lobelia* that he saw on the island and that it was more plentiful over the area where it occurred than perhaps any other lobeliaceous plant occurring in an equal area on the (Hawaiian) group". Members of the Lobeliaceae are characteristic of our rain forests. These records of the *Delissea* from the 1850s and 1860 give some idea of the probable nature of the forest that clothed the mountain of Niihau, before it was decimated by grazing stock.

Lobelia

Lobelia niihauensis St. John, B.P. Bishop Mus., Occas. Papers 9(14): 9, 11, pl. 3, 1931.

Kaali, Jan. 1912, *J.F.G. Stokes* (holotype of the species, in the Bishop Museum); Kaalipuaa, steep basalt cliff, 800 ft. alt., *St. John 23,575*. The writer found it as a low shrub, with its branches hugging the basalt cliff. It grew in inaccessible crevices and small branches were obtained by throwing stones at them.

Goodeniaceae (Naupaka Family)

Scaevola

Low, prostrate shrub; leaves 1-1½ in. long; calyx cylindric, unlobed. *S. coriacea*

Erect shrub, 3-6 ft. tall; leaves 3-5 in. long; calyx with 5 lobes. *S. taccada* var. *sericea*

Scaevola coriacea Nutt.

"Niihau", 1851-55, *Rémy 315 bis*. This was listed also by Hillebrand (1888: 266), and by Skottsberg (1927: 31), where he lists the number erroneously as 3151. It was entered in the Rémy number list at Paris as "315 bis, Scaevola, Nihau". The specimen has not been seen by the writer, but as it was verified both by Hillebrand and by Skottsberg, there is no reason to doubt its identity. It is a maritime littoral species, endemic to the Hawaiian Islands, and still persisting there, but very rarely. It is probably extinct on Niihau.

Scaevola taccada var. *sericea* (Vahl) St. John

S. frutescens (Mill.) Krause var. *sericea* (Forst. f.) Merr.
S. koenigii sensu C.N. Forbes, not Vahl.

Naupaka, or *'aupaka*. The name *naupaka* is considered more correct, but *'aupaka* is now generally used by the younger people.

Kii, Jan. 1912, *J.F.G. Stokes*; Poooneone, sand dunes near beach, occasional, 20 ft. alt., *St. John 22,745*. Also recorded by the writer at Kaununui, Kamalino and Kawaihoa Point, but actually it is abundant generally along both coasts.

Compositae (Sunflower Family)

Key to the Genera

Fruits in large burs with hooked spines. *Xanthium*
Fruits not so.
  Flowers in the head all strap-shaped, yellow. *Sonchus*
  Flowers not so.
    Leaves sharply spiny; flowers red. *Cirsium*
    Leaves unarmed.
      Flowers white.

  Outer flowers of head strap-shaped.
    Seed apex naked.   *Eclipta*
    Seed apex hairy.
      Outer bracts of heads dry, scaly. *Gnaphalium*
      Outer bracts not so.
        Bracts of flower head unequal, overlapping.
                      Aster
        Bracts of flower head equal. *Erigeron*
    Flowers all tubular; shrubs.   *Pluchea*
  Flowers not white.
    Flowers all tubular.
      Outer bracts of heads spiny.   *Centaurea*
      Outer bracts not so.
        Flowers blue.
          Leaves alternate.   *Vernonia*
          Leaves opposite.   *Ageratum*
        Flowers not blue.
          Flowers reddish.   *Emilia*
          Flowers not reddish, pale; receptacle naked.
                     Artemisia
    Outer flowers strap-shaped, yellow; receptacle scaly.
      Scales on receptacle flat.   *Bidens*
      Scales on receptacle folded around each flower.
                  Lipochaeta

         Ageratum

Ageratum conyzoides L.
  Maile-hohono
  Kaali, Jan. 1912, *J.F.G. Stokes*. Not collected by the writer, but observed at Kii, Kaali Cliff, Mokouia Valley, Haao Valley, Apana Valley, Halalii Lake and Poooneone. It is a weed, widely established and introduced from tropical America.

         Artemisia

Artemisia australis Less.
  Hinahina (= silvery), in allusion to the white hairy leaves

Kaaliwai, crevices of basalt cliff, abundant at 400 ft. alt., and from there up, *St. John 23,573*; half mile north of Pueo Point, sea cliff, crevices of precipice, 1,200 ft. alt., *St. John 22,799*; foot of plateau, S. E. Jan. 1912, *J.F.G. Stokes*. Also observed at first valley west of Kaali Cliff and at Paniau. Locally abundant on steep, rocky places.

Aster

Aster sandwicensis (A. Gray) Heiron.
Puuwai, flats and mounds near mudflats, March 12, 1977, *C. Christensen 172*. An annual endemic to Hawaii.

Bidens

Bidens pilosa (L.) var. *minor* (Bl.) Sherff
Ki; ko'oko'olau
Kaali, Jan. 1912, *J.F. G. Stokes*. Also observed by the writer at Kii, Kaali, Kawaewae, Kamalino and Kawaihoa Point. It is a weed introduced from America.

Bidens stokesii Sherff, Bot. Gaz. 70: 101, pl. 12, figs. g-o, 1920; Field Mus. Nat. Hist. Bot. 16(1);142, pl. 19, figs. g-o, 1937.
  B. asplenioides Sherff, Bot. Gaz. 700: 101, pl. 12, figs. a-f, 1920; Field Mus. Nat. Hist., Bot. 16(1): 113, pl. 19, figs. a-f, 1937.
  Campylotheca micrantha sensu C. N. Forbes, not of (Gaud.) Cassini.
Ko'oko'olau
Kaali, Jan. 1912, *J. F. G. Stokes;* Kaaliwai, crevices in basalt cliff, 700 ft. alt., *St. John 23,571;* foot of plateau, southeast, Jan. 1912, *J.F.G. Stokes*.

Two collections of *Bidens* from Niihau were available to Sherff during his revision of the Hawaiian species. The Stokes collection from "foot of Plateau, S. E.," a young flowering branch, had small ovate leaflets and few-headed inflorescences. The one from Kaali, at the opposite end of the mountainous upland was a mature branch with lanceolate, acuminate leaflets and a many-headed inflorescence, past maturity, with the fruit mostly shed. The two appear different. In 1949 this writer found a good stand of *Bidens* in a cliff crevice, 100 feet below the spring at Kaaliwai. The plants were one meter tall, bushy, branched with numerous strong lateral branches. The collection *St.*

John 23,571 was made from several plants and their lateral branches and they show good foliage, flowers, and fruit. From Sherff's original descriptions and from his monograph, the differences that he stated between *B. asplenioides* and *B. stokesii* have been compiled. They are numerous; consisting of petiole length; leaflet shape, toothing, and apex; peduncle length; numbers of heads; outer involucral bract number and length; and achene length. Using the two type collections and one isotype of one species, two of the other, and the recent full collection, *St. John 23,571*, these species have been compared and their characters reviewed. The stated characters of petiole, peduncle, involucral bracts, heads and achenes all merge and disappear when the more abundant material is studied. The leaf shape and apex is the most striking character, but the smaller lower leaves tend to have the leaflets more ovate, while the larger median leaves are more lanceolate. It is the conclusion of the writer that these three collection represent only one population and one species. *Bidens asplenioides* is here reduced to synonymy, and the name *B. stokesii* is selected in preference since it honors the scientist, J.F.G. Stokes, who made the first good collection of plants on Niihau.

Centaurea

Centaurea melitensis L.
   Mokouia Valley, opening in *Prosopis* grove, weed by path, 300 ft. alt., *St. John 23,594*. A weed introduced from the Mediterranean region.

Cirsium

Cirsium vulgare (Savi) Ten.
   C. lanceolatum (L.) Hill
   Pua poni (= a flower, purple)
   Observed at Kapaka Valley and Kaeo.

Eclipta

Eclipta alba (L.) Hassk.
 E. prostrata (L.) L.
 Pua o ka lani
 South shore, Halalii Lake, wet grassy marshy spot 20 ft. alt., *St. John 23,653*. A weed introduced from tropical America.

Emilia

Emilia javanica (Burm. f.) C. B. Robins.
 E. sonchifolia sensu St. John & Hosaka, not of (L.) DC.
 Pua lele (= flower, jumping)
 Mouth of Kaumuhonu Valley, in open *Prosopis* forest on flat, 50 ft. alt., *St. John 23,631*. A weed introduced from Java.

Erigeron

Erigeron bonariensis L.
 E. albidus (Willd.) A. Gray
 Lani wela (= heaven, hot)
 Observed at Halalii Lake. It is a weed introduced from tropical America.

Erigeron canadensis L.
 E. albidum sensu C. N. Forbes, not of A. Gray.
 Kaaliwai, in grassy vales, occasional, 30 ft. alt., *St. John 23,568*. It is a weed introduced from North America.

Gnaphalium

Gnaphalium sandwicensium Gaud. var. *sandwicensium*
 G. sandwicensium Gaud. var. *typicum* Sherff f. *canum* Sherff and f. *olivaceum* Degener & Sherff
 Pu-heu
 Oiamoi, sand dunes, 20 ft. alt., *St. John 23,651*. According to Article 26 of International Code of Botanical Nomenclature (Seattle, 1972), the name var. *typicum* Sherff must be replaced by var. *sandwicensium* Sherff, and Degener and Sherff distinguished two

formae under this variety, solely on the basis of whether the herbage was more or less whitish (f. *canum*), or whether at least the young growth was olive green (f. *olivaceum*). Both kinds are well represented in the Bishop Museum, but in a number of collections the two are intermingled and apparently they grew together. The two look like growth stages, of ecads, of even conditions resulting from the manner of drying the plants. The writer has decided that the two should be reduced to one. Sherff failed to find Gaudichaud's original type specimen so preferred to give each of the forms a new name.

Lipochaeta

Leaves fleshy, narrowly spatula-shaped, 3-7 mm. wide; heads
  about 12 mm. wide.   *L. integrifolia*
Leaves flat, much broader.
  Blades mostly without petioles.   *L. succulenta*
  Blades with petioles.
    Blades heart shaped.   *L. niihauensis*
    Blades not heart-shaped.
      Blades not lobed; heads 2.5 cm. across.   *L. kawaihoaensis*
      Blades 3-lobed; head 2 cm. or less across.
        Sides lobes of blades egg-shaped.
                                L. lobata var. *maunaloensis*
        Sides lobes of blades lance-shaped, sharp pointed.
                                      L. lobata var. *incisior*

Lipochaeta integrifolia (Nutt.) A. Gray var. *integrifolia*
    North Kona coast, Jan. 1912, *J.F.G. Stokes*. (This was determined and labelled by C.N. Forbes as *Nama sandwicensis,* but it is a flowering specimen of *Lipochaeta!*) First bay north of Kawaihoa Cone, sprawling over stunted *kiawe* shrubs, 25 ft. alt., Jan. 15, 1977, *C. Christensen 111*; Pali koa'e, sandy soil near shore, Feb. 12, 1977, *C. Christensen 146*.

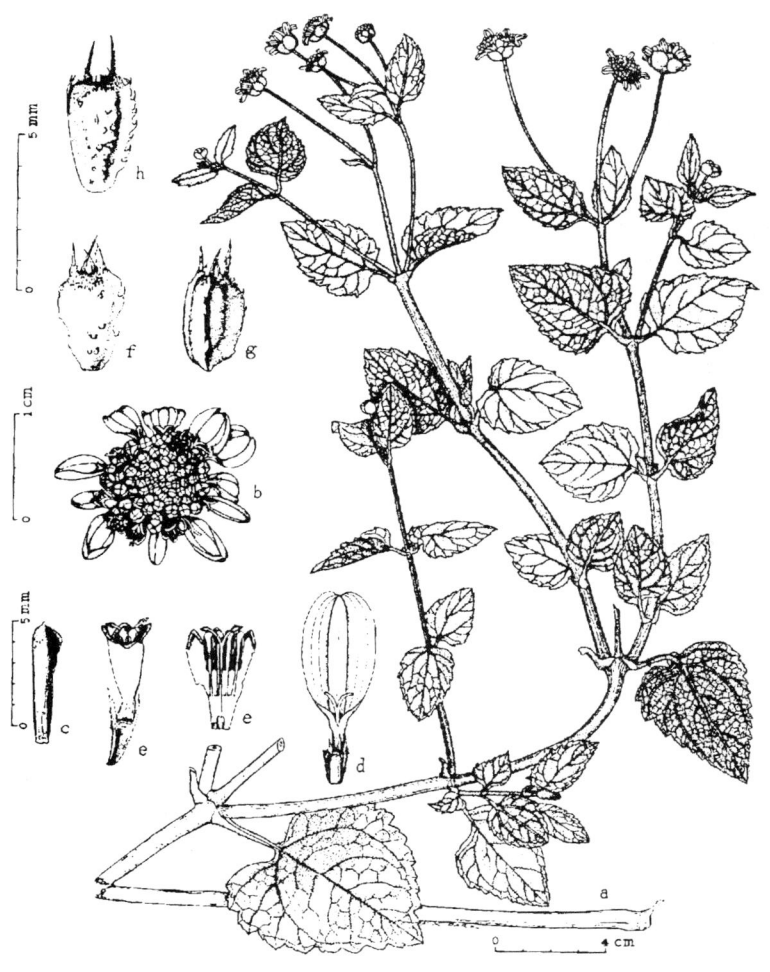

Lipochaeta niihauensis, from holotype: *a*, habit X 1/2; *b*, head X 2; *c*, chaff X 4; *d*, ray flower X 4; *e*, disk Flower X 4; *f*, ray achene, dorsal view, X 6; *g*, ray achene, lateral view, X 6; *h*, disk achene, dorsal view, X 6.

Lipochaeta kawaihoaensis St. John, Pacif. Sci. 13: 181-182, 184-185, fig. 9. 1959.
  Koʻokoʻolau
  Kawaihoa Point, Steep gully, 300 ft. alt., St. John 23,611. This is the type and only known collection of this species which is endemic to Niihau.

Lipochaeta lobata D.C. var. *incisior* St. John, Pacif. Sci. 13: 185, Fig. 10. 1959.
  Nehe
  Kaaliwai, in thicket with *Euphorbia celastroides* and *Artemisia australis*, 750 ft. alt., St. John 23,572, the type specimen of this variety which is endemic to Niihau.

Lipochaeta lobata DC. var. *maunaloensis* Sherff
  South half of island, Jan. 1912, *J. F. G. Stokes*. This matches well the original specimen of this variety which was described from Molokai.

Lipochaeta niihauensis St. John, Pacif. Sci. 13: 186, 188-189, fig. 11. 1959.
  Paʻapaʻaʻina (= to crinkle, probably in allusion to the stiff, brittle leaf blades)
  Kii, among rocks on basalt knoll, 100 ft. alt., *St. John 23,664*. This is the type collection and the species is endemic to Niihau.

Lipochaeta succulenta (Hook. & Arn.) DC. var. *succulenta*
  The type of this species was collected during the Beechey Voyage by G.T. Lay and A. Collie on June 1-2, 1826, "among volcanic rocks on the shore of the island of Oneeheow (Niihau), where it is called *Nehe* or *Nenehe*," according to Hooker and Arnott (1830-1841: 87). It was also collected in 1851-1855 on Niihau, *J. Rémy* 257, and this record was published by Hillebrand (1888: 207) and by Sherff (1935: 38-40), who verified the determination. There is also a small duplicate of it in the B. P. Bishop Museum. Sherff also studied the type specimen of the species and reported (1935: 40), "The type sheet at Kew is from Hooker's private herbarium, and its inscription says

'Oahu'. Since, however, the species is not otherwise known to me from Oahu, it may be that this citation is erroneous. Or can it be that specimens were collected on both islands, the one from Oahu escaping the notice of the authors originally while the on from Niihau failed to be preserved to the present day."

Sherff's second suggestion is a possibility, but it does not seem a probability. The *H.M.S. Blossom*, under Capt. F.W. Beechey, visited both islands, and the naturalists collected plants on both Niihau and Oahu. The occurrence of *L. succulenta* on Niihau was confirmed by its rediscovery there, in 1851-1855 by J. Rémy. It is also known from Kauai, where it was collected by J. Rémy, and more recently by U. Faurie, and in 1948 by A.M. Alexander and L. Kellogg. It is a succulent, littoral halophyte. The indigenous or endemic littoral species are still well preserved in the Hawaiian Islands, even on the densely populated island of Oahu. Incidentally, the flora of Oahu is best known of that of any of the islands, and none of the many botanists to study its flora since 1826 have collected the species there. It seems evident that the type collection by Lay and Collie was from Niihau, as originally published.

Pluchea

Pluchea indica (L.) Less.
   Liua
   South shore of Halalii Lake, on dry salt flats, 50 ft. alt., *St. John 22,735*. Also observed at Halulu Lake and Kawaewae. It is abundant on salt flats. It is a weed introduced from tropical Asia.

Pluchea odorata (L.) Cass.
   Liua
   Halalii Lake, south border, subsaline or arid flats, 50- ft. alt., *St. John 22,737*. It is a weed introduced from tropical America, arriving on Niihau about 1937.

Sonchus

Sonchus oleraceus L.
Pua lele (= flowering, jumping)
Kii, weed in corral, 15 ft. alt., *St. John 23,581*. Also observed at Mokouia Valley, Halalii Lake and Kawaewae. It is a weed introduced from Europe.

Vernonia

Vernonia cinerea (L.) Less.
Mokouia Valley, under basalt ledge, 700 ft. alt., *St. John 23,590*. It is a weed introduced from the tropics of the Old World.

Xanthium

Xanthium sp.
Kikania
North of Halulu Lake, on mud flats, *St. John 23,645*. It is a weed introduced from North America. The plants collected were very juvenile.

Select Bibliography

Bailey, F. Manson. 1900. *Queensland Flora* 2: 1-737. H.J. Diddams & Co., Brisbane.

Beaglehole, J.C. 1967. *The Journals of Captain Cook on his Voyages of Discovery. The Voyage of the Resolution and Discovery, 1776-1780* 2: 723-1,647.

Beechey, F.W. 1831. *Narrative of a Voyage to the Pacific and Beering's Strait, in H.M.S. Blossom* 1: xxi, and 1-742. Henry Colburn & Richard Bentley, London.

Broughton, William Robert. 1804. *A Voyage of Discovery to the North Pacific Ocean in the Providence, 1795, 1796, 1798*, xx, and 1-393.

Burkill, Isaac Henry. 1935. *A Dictionary of the Economic Products of the Malay Peninsula* 1: xi, and 1-1,220. Crown Agent from the Colonies, London.

De Candolle, Auguste Pyramus. 1825. *Prodromus Systematis Natualis Regni Vegetabilis* 2: 1-644. Treuttel et Wiirtz, Paris.

Colnett, James. 1788. *Journal of the Voyage of the Prince of Wales*. Manuscript in Archives of Hawaii. (He was in the Sandwich Islands, January 1 to March 20, 1788.)

Forbes, Charles Noyes. 1913. "An Enumeration of Niihau Plants". *Ocas. Pap. Bernice P. Bishop Mus.* 5(3): 17-26, 3 pl. 1 map.

Handy, E.S. Craighill. 1940. "Hawaiian Planter, Vol. 1, His Plants Methods and Areas of Cultivation". *Bernice P. Bishop Mus. Bull.* 161: iii, and 1-227, 8 pl.

Heller, A.A. 1897. "Observations on the ferns and flowering plants of the Hawaiian Islands". *Minnesota Geol. & Nat. Hist. Survey Bulletin* 9; *Minn. Botanical Studies* 1: 760-922, pl. 42-69.

Hillebrand, William. 1888. *Flora of the Hawaiian Islands*, xcvi, and 1-673, frontispiece, 4 maps. Carl Winter, Heidelberg.

Hooker, Joseph Dalton. 1879. *The Flora of the British India* 2: 1-792. L. Reeve & Co., London.

Hooker, William Jackson, and G.A. Walker Arnott. 1830-1841. *The Botany of Captain Beechey's Voyage, H.M.S. Blossom*, 1-486. Treuttel & Würtz, London.

Hosaka, Edward Yataro. 1936. "A Troublesome Introduced Grass". *Mid-Pacific Magazine* 49: 126, 1 fig.

Judd, Charles Sheldon. 1920. "The Wiliwili Tree". *Hawaiian Forester and Agriculturist* 17(4): 95-97, 2 pl.

Judd, Charles Sheldon. 1932. "Niihau". *Hawaiian Forester and Agriculturist* 29: 1-9.

Krukoff, B.A. 1939. "Preliminary Notes on Asiatic-Polynesian Species of *Erythrina*". *Journal of the Arnold Arboretum* 20: 225-233.

Lamarck, Jean Baptiste. 1789. *Encyclopédie Méthodique Botanique* 3: viii, and 1-759. Panckoucke, Paris.

Lecomte, Henri. 1908-1923. *Flore Générale de L'Indo-Chine* 2: 1-1,212. Masson et Cie., Paris.

Mann, Horace (Jr.). 1866-1871. "Flora of the Hawaiian Islands". *Essex Inst., Communic.* 5: 113-144 (1866); 161-176 (1867); 177-192 (1867); 233-248 (1868); 6: 105-112 (1871).

Mann, Horace (Jr.). 1867. "Enumeration of Hawaiian Plants". *Proc. Amer. Acad. Arts* 7: 143-235.

Munro, George C. 1952. "Attempts to save the shoreside and dryland plants of Hawaii". *Elepaio* 13(1): 1-5.

Pollacci, Gino. 1914. "Sull. 'Abrus precatorius' ". *L. R. Inst. Botanco Universita Pavia* 15: 285-290, t. xviii.

Portlock, Nathaniel. 1789. *A Voyage Round the World, but more particularly to the North-West Coast of America; performed in 1785, 1786, 1787, and 1788, in the King George and Queen Charlotte, Captains Portlock and Dixon*, xii, and 1-382, x1, 20 figs. Stockdale and George Goulding, London.

Pukui, Kawena. 1933. "Hawaiian Folk Tales". *Folklore Publ.* Ser. 3, 13: 127-185. Vassar College, Poughkeepsie.

Rock, Joseph Francis. 1913. "The Indigenous Trees of the Hawaiian Islands". *Hawaiian Gazette* vi, and 1-518, pl. 1-215. Honolulu.

Rock, Joseph Francis. 1919. "The Arborescent Indigenous Legumes of Hawaii". *Botanical Bulletin* Territory of Hawaii, Board of Agriculture and Forestry, Division of Forestry 5: 1-53.

Rock, Joseph Francis. 1919. "A Monographic Study of the Hawaiian Species of the tribe Lobelioideae, Family Campanulaceae". *Mem. Bernice P. Bishop Mus.* 7(2): XVI, and 1-395, pl. 1-217.

Rock, Joseph Francis. 1920. *The Leguminous Plants of Hawaii*, x, and 1-234, pl. 1-93. Hawaiian Sugar Planter's Association Experiment Station, Honolulu.

Roxburgh, William. William Carey ed. 1832 (and reprint 1874, C.B. Clarke ed.). *Flora Indica or Descriptions of Indian Plants*, vii, and 1-763, lxiv. Thacker, Spink & Co., Calcutta.

St. John, Harold. 1948. "Report on the Flora of Pingelap Atoll, Caroline Islands, Micronesia, and Observations on the Vocabulary of the Native Inhabitants. Pacific Plant Studies 7. *Pacific Sci.* 2(2): 96-113, figs. 1-9.

St. John, Harold. 1952. "A New Variety of *Pandanus* and a New Species of *Fimbristylis* from the Central Pacific Islands". *Pacific Plant Studies 11. Pacific Sci.* 6(2): 145-150, figs. 1-3.

St. John, Harold. 1959. "Botanical Novelties on the Island of Niihau, Hawaiian Islands". *Hawaiian Plant Studies 25. Pacific Sci.* 13: 156-190, figs. 1-11.

St. John, Harold. 1969. "Monograph of the genus Brighamia (Lobeliaceae)". *Hawaiian Plant Studies 29. J. Linn. Soc., Bot.* 62: 187-204, 7 figs., 2 pl.

Samwell, David. 1967. "Some Account of a Voyage to South Seas, In 1776-1777-1778", published by Beaglehole, J.C., in the *Journals of Captain James Cook on his Voyage of Discovery. The Voyage of the Resolution and Discovery, 1776-1780* 3(2): 989-1,300.

Scanlan, Grace Margaret. 1942. "A Study of the Genus Cyerus in the Hawaiian Islands". *Catholic University American, Biol. Ser.* 41: viii and 1-62.

Sherff, Earl Edward. 1935. "Revision of Tetramolopium, Lipochaeta, Dubautia, and Railliardia". *Bernice P. Bishop Mus. Bull.* 135: 1-136, figs. 1-43.

Sherff, Earl Edward. 1938. "Revision of the Hawaiian Species of Euphorbia L". *Ann. Missouri Bot. Gard.* 25: 1-94.

Sherff, Earl Edward. 1945. "Revision of the genus Schiedea Cham. & Schlecht.". *Brittonia* 5(3): 308-336.

Sherff, Earl Edward. 1951. "A revision of the Hawaiian Island genus Nototrichium Hillebrand, (Amaranthaceae)". *Bot. Leaf..* 4: 2-21.

Sherff, Earl Edward. 1952. "Further studies of Hawaiian Araliaceae: Additions to Cheirodendron Helleri Sherff and a preliminary treatment of the endemic species of Reynoldsia A. Gray". *Bot. Leaf.* 6: 6-19.

Simmonds, J.H. 1927. *Trees from other lands for shelter and lumber in New Zealand*, xviii, and 1-164, pl. 1-74. Brett Printing and Publishing Company, Auckland.

Sinclair, Mrs. Francis. 1885. *Indigenous Flowers of the Hawaiian Islands*, vii, and pl. 1-44, with text. Sampson, Low, Marston, Searle, and Rivington, London.

Skottsberg, C. 1927. "Artemisia, Scaevola, Santalum, and Vaccinium of Hawaii". *Bernice P. Bishop Mus. Bull.* 43: 1-89, pl. 1-vii, figs. 1-30.

Stearns, Harold T., and Gordon A. Macdonald. 1947. "Geology and Ground-Water Resources of the Island of Niihau, Hawaii; and Petrography of Niihau". *Bull.* Territory of Hawaii, Division of Hydrography 12: 1-53, pl. 1-58, figs. 1-8, map.

Trimen, Henry. 1894. *A Hand-Book to the Flora of Ceylon* 2: 1-392, Dulau and Co., London.

Turnbull, John. 1813. *A Voyage Round the World in the years 1800, 1801, 1802, 1803, and 1804*, ed. 2 xv, and 1-516. A. Maxwell, London.

Watt, George. 1889. *A Dictionary of the Economic Products of India* 1: xxxiii, and 1-559. Government Printing Office, Calcutta.

Whitney, Leo D., F.A.I. Bowers & M. Takahashi. 1939. "Taro Varieties in Hawaii". *Hawaii Agric. Exp. Sta. Bull.* 84: 1-86, figs. 1-6.

Whitney, Leo D., and Edward Y. Hosaka. 1936. "New Species of Hawaiian Panicum and Eragrostis". *Ocas. Pap. Bernice P. Bishop Mus.* 12(5): 1-6, figs. 1-2.

Index of Hawaiian Plant Names

An asterisk (*) marks 25 names used on Niihau which are applied to other species on other islands. See also St. John "Vernacular Plant Names Used on Niihau Island", *Occasional Papers of Bernice P. Bishop Museum* Vol. XXV No. 3:19.

ʻAʻaliʻi	107	ʻIhi laʻau	52
ʻAeʻae	131,132	Ilima laukahi	110
ʻAeʻae mahu	87	Ilima lau liʻiliʻi	110
ʻAhuʻawa	66	Ilima pua kea	109
ʻAi-a-ka-manu	135	ʻInia	102
ʻĀkia	132	Inikoa	96
ʻAkiʻaki	103,104,105	ʻIwaʻiwa	52
ʻAkoko	87,88	Kaʻa *	68
ʻĀkulikuli	79	Kaʻaliwai	142
ʻAlaʻalawai nui	52	Kai ʻoʻio	62
Alamoho *	101	Kākaliaoa	93
ʻAlani	101	Kalamoho *	52
ʻAlani Hawaii	102	Kaunoʻa	89
ʻAlani pakē	84	Kaunoʻa-lei	121
Aloalo	109	Kāwelu	59
Anena	86	Ki	140
ʻAupaka	138	Kiawe	98
ʻEhuʻawa	66	Ki hohono	130
Hākonakona	63	Kīkania	147
Hala *	53,74	Kinikini	93
Hale o ka iʻa	107	Kō	64
Hau	109	Koa *	97
Hau hele *	108	Koali	125
Haʻu-o-i	129	Koali lāʻau	125
Haʻu-o-i- pakē	110	Koa mahu	97
Hāwane	72	Kohekohe	68
Heʻi	113	Kolī	106
Heʻu pueo	62	Kolokolo	108,129
Hiʻaloa	112	Kolomona *	94,95
Hilahila	98	Koʻokoʻolau	140,145
Hinahina	139	Kou	127
Hinahina kahakai	126	Kou haole	127
Hola	100	Kukui	103
Hō-na	110	Kukui pakē	105
Huelo-popoki	56	Kula pepeiao *	109
Hunakai	125	Kupukupu	
ʻIhiʻihi-i-one	90	(see also palapalai)	52

152

Lā'i	75	'Ohai *	
Lanakana	129	(see also 'ohai a Papiahuli)	93
Laniwai	112	'Ohai a Papiahuli	100
Lani wela *	142	'Ohe	65
Lau 'au kepani	83	'Ohe'ohe	116
Lau ehu	62	'Ōhi'a *	131
Lauhihi	118	'Oka	83
Lau-kāhi *	117	'Oka pua 'ula'ula	83
Lau po'ohina	128	'Olena	95
Lihilihi kakahiaka	105	Oliwa haole	120
Lili'u	120	'Olulo	135
Limu-alolo	53	Opiuma	98
Limu-pa-kai	54	Pa'apa'aina	145
Liua	146	Pa'i'iha	52
Mahiki *	65	Paina	79
Mai'a a Maui	90	Paina kahakai	112
Maile-hohono	139	Paina pupupu	112
Makahala *	133	Paka	132
Makaloa	66	Pakapakai	84,85
Malina *	76	Palama	115
Malina haole *	76	Palama waiū	118,119
Manakō	106	Palapalai *	52
Mānienie	58	Palea	135
Mānienie mahiki	58	Pale-piwa	113
Ma'o	109	Pa nini	115
Ma'o (see also hau-hele)	108	Pā pipi	113
Mau'u 'aki'aki	70	Pa'u a Hi'iaka	126
Mau'u alolo	59	Paunu	93
Mau'u haole	55	Piku	83
Mau'u pulumi	62	Pili	60
Miki palaoa	94	Pilipili	95
Milo	111	Pilipili 'ula	95
Moa	52	Pilo *	90
Momona *	119	Pipi wai *	68
Naio	133	Pōhuehue	124
Naupaka	138	Pomelaiki	114
Nehe	145	Po'opo'ohina	128
Neke	71	Pōpolo	132
Ni'ani'au	52	Pua hilahila	98
Nioi	131	Puakala	89
Niu	71	Pua lele	142,147
Niu kahiki	72	Pua o ka lani	142
Nohonoho pu'uone	127	Pua-o-kama	135
Nohu	101	Pua pepa	86
Noni	133	Puapili	90
Nu'umela	120	Pua 'ula'ula	141

Pu aupaka	63	Ūhini	94
Puawa	135	Uhiuhi *	98
Pu-heu	116	ʻUlu	80
Pūkiawe lenalena	93	ʻUme alu	56
Pūkiawe ʻulaʻula	92	Wāhane	72
Pulupulu	109	Waiawī	116
Ti kepanī	94	Wauke	82
ʻUala	122	Wiliwili	96
Uhaloa	112	Wiliwili lenalena	96
Uhi	76		

Index of Families and Genera

Abrus	92	Bignoniaceae	133
Abutilon	108	*Boerhavia*	86
Adiantum	52	Boraginaceae	127
Agave	75	*Bougainvillea*	86
Ageratum	139	*Brighamia*	135
Aizoaceae	87	Bromeliaceae	74
Albizzia	93	*Broussonetia*	82
Aleurites	45, 103	*Byronia*	41
Amaranthaceae	85	Cactaceae	113
Amaranthus	85	*Caesalpinia*	93
Amaryllidaceae	75	*Calotropis*	120
Anacardiaceae	106	*Canavalia*	93
Anacystis	46	Capparaceae	89
Ananas	74	*Capparis*	90
Andropogon	55	*Capsicum*	131
Antidesma	41	*Cardiospermum*	107
Apocynaceae	120	*Carex*	41
Araceae	74	*Carica*	113
Araliaceae	116	Caricaceae	113
Araucaria	53	Caryophyllaceae	88
Araucariaceae	53	*Cassia*	94
Argemone	89	*Cassytha*	89
Artemisia	139	*Casuarina*	79
Artocarpus	80	Casuarinaceae	79
Asclepiadaceae	120	*Catharanthus*	120
Asclepias	120	*Cenchrus*	56
Aster	140	*Centaurea*	141
Atriplex	85	*Ceratonia*	94
Bacopa	132	*Cereus*	113
Batidaceae	87	*Chara*	47
Batis	87	Characeae	47
Bidens	140	*Cheirodendron*	44, 116

Chenopodiaceae	84	*Erythrina*	96
Chenopodium	84	*Eucalyptus*	115
Chloris	57	*Eugenia*	115
Chlorococcum	47	*Euphorbia*	40, 103
Chroococcaceae	46	Euphorbiaceae	102
Chrysophyllum	118	*Ficus*	83
Cirsium	141	*Fimbristylis*	70
Citrullus	134	*Furcraea*	76
Citrus	101	*Gnaphalium*	142
Cocculus	89	Goodeniaceae	138
Cocos	71	*Gossypium*	109
Colocasia	74	Gramineae (Poaceae)	54
Colubrina	108	*Grevillea*	83
Compositae	41, 138	*Gynandropsis*	90
Convolvulaceae	121	*Heliotropium*	127
Cordia	127	*Heteropogon*	60
Cordyline	75	*Hibiscus*	41, 109
Coronopus	90	*Hydrocotyle*	117
Cressa	121	Hydrophyllaceae	126
Crotalaria	94	Hypnaceae	51
Cruciferae	90	*Indigofera*	96
Cucumis	134	*Ipomoea*	121
Cucurbita	134	*Isodendrion*	40, 112
Cucurbitaceae	134	*Isotoma*	40
Cuscuta	121	*Jacquemontia*	126
Cymbopogon	58	*Jasminum*	119
Cynodon	58	*Jatropha*	105
Cyperaceae	65	*Joinvillea*	41
Cyperus	66	Labiatae (Lamiaceae)	129
Delissea	137	*Lantana*	129
Desmodium	95	Lauraceae	89
Digitaria	58	*Lecanora*	48
Dioscorea	76	Lecanoraceae	48
Dioscoreaceae	76	Leguminosae (Fabaceae)	91
Dodonaea	107	*Lepidium*	91
Doryopteris	52	*Leucaena*	97
Dryopteris	52	Liliaceae	75
Duranta	128	*Lipochaeta*	143
Echinochloa	40, 59	*Lobelia*	137
Eclipta	142	Lobeliaceae	135
Ectropothecium	51	*Lycium*	131
Eleocharis	68	*Lycopersicon*	131
Emilia	142	*Lysimachia*	40, 117
Eragrostis	59	*Malva*	110
Erigeron	142	Malvaceae	108
Eriochloa	60	*Malvastrum*	110

Mangifera	106	Physciaceae	49
Manilkara	119	Piperaceae	79
Marsilea	52	*Pithecellobium*	98
Melia	102	*Pityrogramma*	52
Meliaceae	102	*Plectranthus*	130
Melinis	60	*Pluchea*	146
Menispermaceae	89	Plumbaginaceae	118
Merremia	126	*Plumbago*	118
Messerschmidia	128	*Portulaca*	87
Mimosa	98	Portulacaceae	87
Momordica	135	*Potamogeton*	53
Moraceae	80	Potamogetonaceae	53
Morinda	133	Pottiaceae	51
Mucuna	45	Primulaceae	117
Murraya	102	*Pritchardia*	72
Musa	79	*Prosopis*	98
Musaceae	79	Proteaceae	83
Myoporaceae	133	Protococcaceae	47
Myoporum	133	*Psidium*	116
Myrtaceae	114	*Psilotum*	52
Nama	126	*Psychotria*	41
Nephrolepis	52	*Puccinia*	47
Nicotiana	132	Pucciniaceae	47
Noronhia	119	*Punica*	114
Nototrichium	86	Punicaceae	114
Nyctaginaceae	86	*Ramalina*	48
Ocimum	130	*Reynoldsia*	44, 116
Olea	120	Rhamnaceae	108
Oleaceae	119	*Rhynchelytrum*	63
Opuntia	113	*Ricinus*	106
Oxalidaceae	101	*Rosa*	91
Oxalis	101	Rosaceae	91
Palmae	71	Rubiaceae	133
Pandanaceae	53	*Ruppia*	54
Pandanus	53	Ruppiaceae	54
Panicum	60	Rutaceae	101
Papaveraceae	89	*Saccharum*	63
Parmelia	48	*Salvia*	130
Parmeliaceae	48	*Sansevieria*	75
Paspalidium	63	Santalaceae	84
Passiflora	112	*Santalum*	84
Passifloraceae	112	Sapindaceae	106
Peperomia	79	Sapotaceae	118
Phaseolus	98	*Scaevola*	40, 138
Phoenix	72	*Schiedea*	41, 88
Physcia	49	*Schinus*	106

Schizostachyum	65	*Tecomaria*	133
Scirpus	70	*Telochistes*	50
Scrophulariaceae	132	Teloschistaceae	50
Scytonema	46	*Tephrosia*	100
Scytonemataceae	47	*Thespesia*	111
Selaginella	41	*Tribulus*	101
Sesbania	100	Umbelliferae	117
Sesuvium	87	Usneaceae	48
Setaria	65	*Verbena*	129
Sicyos	135	Verbenaceae	128
Sida	110	*Vernonia*	147
Solanaceae	130	Violaceae	112
Solanum	132	*Vitex*	40, 129
Sonchus	147	*Waltheria*	112
Sporobolus	65	*Weisia*	51
Sterculiaceae	112	*Wikstroemia*	41
Strongylodon	45	*Xanthium*	147
Tamaricaceae	112	Zygophyllaceae	101
Tamarix	112